# CSR, Sustainability, Ethics & Governance

**Series Editors**

Samuel O. Idowu, London Metropolitan University, Calcutta House, London, UK

René Schmidpeter, Cologne Business School, Cologne, Germany

In recent years the discussion concerning the relation between business and society has made immense strides. This has in turn led to a broad academic and practical discussion on innovative management concepts, such as Corporate Social Responsibility, Corporate Governance and Sustainability Management. This series offers a comprehensive overview of the latest theoretical and empirical research and provides sound concepts for sustainable business strategies. In order to do so, it combines the insights of leading researchers and thinkers in the fields of management theory and the social sciences – and from all over the world, thus contributing to the interdisciplinary and intercultural discussion on the role of business in society. The underlying intention of this series is to help solve the world's most challenging problems by developing new management concepts that create value for business and society alike. In order to support those managers, researchers and students who are pursuing sustainable business approaches for our common future, the series offers them access to cutting-edge management approaches.

*CSR, Sustainability, Ethics & Governance* is accepted by the Norwegian Register for Scientific Journals, Series and Publishers, maintained and operated by the Norwegian Social Science Data Services (NSD)

Paolo Candio

# Sustainability and Corporate Performance in Health Care

## ESG Implications for the European Industry

Springer

Paolo Candio
University of Trento
Trento, Italy

ISSN 2196-7075 ISSN 2196-7083 (electronic)
CSR, Sustainability, Ethics & Governance
ISBN 978-3-031-63984-5 ISBN 978-3-031-63985-2 (eBook)
https://doi.org/10.1007/978-3-031-63985-2



This Springer imprint is published by the registered company Springer Nature Switzerland AG
The registered company address is: Gewerbestrasse 11, 6330 Cham, Switzerland

# Foreword

The relationship of the company with its environment, understood as the social, economic, natural, political, and competitive context, is not a subject of study separate from that of the company itself. The company expresses its deepest nature precisely in the relationship between its internal dynamics—the economic combinations and coordination, as Gino Zappa would say—and the world of which it is a part, contributing to its evolution. The theme of sustainability, in its broadest sense as an attribute of an activity capable of satisfying current needs without compromising the future ability to satisfy those of future generations, is inherent in the very essence of the company, which is, by definition, an "institution intended to satisfy human needs in order to endure." Such "durability" obviously depends on the durability of the environment in which the company operates. The fact that only in recent years the theme of sustainability has become so prominent in everyone's life, including that of companies, does not mean that it was previously irrelevant. The company system is naturally an open system that physiologically poses the problem of maintaining economic conditions over time as a condition of its own survival. However, the same concept of economic viability that should characterize business management is defined in broader terms than just the momentary balance between costs and revenues: it is necessarily intended as the maintenance over time of the conditions of the company's existence.

If political economy teaches us to understand the sense of externalities, that is, the effects of actions that do not directly affect the company's economy, business economics investigates the internal and external conditions for maintaining balance without separating external aspects from internal ones, because all contribute to achieving the requirement of sustainability. Therefore, all are inevitably of interest to business management. Understanding management phenomena allows us to grasp the underlying reasons for fundamental processes, to understand their effects. Adopting a systemic view, the company is seen as an element in strong mutual relationship with other elements, namely with its environment: a system in which the boundaries between company and environment are blurred and changing over time. The principle of sustainability, understood precisely as the ability to maintain equilibrium over time, therefore has a broader scope and greater implications in terms of management than the goal of profit characterizing studies on business economics subject to the axiom of maximization. A maximization that obviously disregards what is not directly and immediately attributable to the costs and revenues of the company. This relationship with the environment has assumed in recent theoretical construction the connotation of social responsibility, that is, the integration of social and environmental considerations and relations with various stakeholders in management decision-making. A wider array of stakeholders has seen inclusion in decision-making processes over time, driven by the evolution of social sensitivity to the role of the company and its impacts while considering the limited natural resources available, different categories of subjects affected by its activities, ultimately including future generations.

However, business economics also has its normative purpose, aiming to provide rules of conduct and tools aimed at governing the company and managing resources. Differing from other social sciences, business science does not separate the observation and interpretation of reality and action on clearly distinct levels. The development of knowledge supporting management processes and the appreciation of the results derived from them are indeed two sides of the same coin. If the technique of the balance sheet has generated the reference model for the definition of the notion of business economic balance, business economics itself makes the development of balance sheet techniques and rules one of its fundamental objects of study. An abstract concept (economic balance) and a measurement technique (techniques and principles of financial statement construction) are not in this case in a hierarchical sequence, but in mutual interrelation. It is therefore evident that to describe how a company takes into account its financial, environmental, and social impacts on decisions, the dimension of social responsibility assumes great importance.

Managers, consumers, investors, workers, political entities have increasingly understood how economic growth is closely linked to social and environmental well-being, not only because social and environmental well-being are fundamental "goods" for people's lives, but because they are the prerequisite for maintaining equilibrium over time, a cornerstone concept of business management economics. Although developing awareness of corporate social responsibility is predominantly a voluntary action, there is also increasing pressure on companies to make a positive contribution to the communities in which they operate, or at least to reduce their negative impact. But this pressure is not an external interference or demand; it is the natural manifestation of a need that is also that of the company, which cannot do without it because it is a condition for its continuity over time. And it is therefore the primary interest of companies to serve this function.

At the international level, governments have initiated a process of concrete implementation of some key elements that give meaning to corporate social responsibility, with a great focus on the concept of sustainability. Recalling how measurement is a constituent element of business results, attention must be paid to the ability of "responsible" companies (i.e., sensitive to their social and environmental role) to measure the (positive or negative) impact that their behavior generates. However, it is plausible that "less responsible" companies are more attentive to the negative impact that any of their "responsible" decisions would have on their profits, emphasizing short-term results at the expense of their ability to maintain them over time. Corporate sustainability itself, on the other hand, is now and in the future, strongly correlated with companies' ability to take into account the environmental, social, and governance consequences of their decisions and actions. Clearly, forward-thinking companies that want to be particularly attentive to such consequences and be genuinely responsible must face many challenges and overcome a series of barriers. Among these, the incentive systems for managers, always based on measured results, certainly play an important role. It is therefore necessary to strengthen the ability to evaluate all the advantages and disadvantages associated with social responsibility and set short-, medium-, and long-term goals for achieving corporate sustainability. The concept of sustainability applied in the context of governance provides the necessary elements not only for the development of corporate strategy but can also act as a catalyst for it. In other words, if corporate governance is also guided by respect for the principle of sustainability, it will be easier to define a framework useful for laying the foundations of cultural reinvention within a corporate context.

Health care companies, the subject of this work, who integrate the principle of sustainability into their priority objectives, can expand not only the range of programs that the company itself can boast but also and above all the possible healthcare services they offer to the population. The contributions currently present in the literature relating to management systems implemented in health care companies highlight the existence of a variety of key factors that directly and indirectly influence corporate sustainability. It has already been demonstrated how management expenses often have a (perhaps obvious) direct and positive relationship with the performance level of healthcare companies in areas related to social responsibility. Therefore, economic interpretation based solely on management costs is not able to guide corporate decisions towards sustainability. Operating within a framework where the constraints are the balance between costs and revenues and minimum service levels certainly does not allow expanding the perspective to dimensions that, while fundamental in a long-term perspective, escape every representation.

Representation models of business facts and their consequences are indeed fundamental for guiding company governance and management. They allow conscious decision-making, guiding behaviors through incentives, correcting errors, perceiving new emergencies, and ultimately achieving accountability towards stakeholders that directs the organization's actions. Data collection, organization, and management are, as business economics teaches us, inseparable elements. On the other hand, the reorganization that has affected health care service networks and patient care methods in recent years has made it possible to define new paths extended to the sustainable development of health care companies, whose socio-environmental impacts must be appreciated.

It is precisely on these aspects that Candio's work focuses, rigorously tracing the development of literature and regulations regarding non-financial disclosure to sustainability reporting and delving into the evolutionary analysis of financial and sustainability performance of listed companies for the European context and where improvements in ESG profile reward companies in terms of corporate performance.

The issue of evaluating sustainability-related investments in health care companies is a current topic, which, also spurred by recent national and European policies, is certainly of great relevance. Thanks to an original quantitative analysis, Candio's work also allows policymakers and operators to provide a useful basis for making informed decisions regarding sustainability-related investments in the company, illustrating what aspects and skills to develop in order to orient the culture and organizational processes towards sustainability objectives and improved corporate performance. In this sense, the author's research aligns with the idea of business economics as both a theoretical science, attentive to the study of reality, and a normative one, oriented to the study of what reality should be, providing updated ideas, models, and tools. With the vast availability of information today and thanks to the management applications introduced even in health care company management processes, it is possible to include sustainable development data and information in accountability in order to initiate a virtuous circle between information and decisions, allowing health care companies to be real architects of a sustainable policy for the environment and the community and therefore ultimately for themselves. The detailed and exemplary dimension of the topics adopted by the author allows understanding the processes that health care companies can implement in order to pursue the principle of sustainability in the most effective and efficient way possible, thus realizing the fundamental prerequisite of the "durability" of their own organization which, in order to survive, must adapt to changes in its environment and be, at the same time, its architect.

Department of Economics and Management Paolo Collini
University of Trento,
Trento, Italy
12 April 2024

# Acknowledgements

I am deeply indebted to my family, who have borne the brunt of my absence.

# Introduction

In recent decades, there has been increasing scrutiny of the traditional model of endless growth, with sustainability gaining prominence as a key concept. This shift has led to a reevaluation of corporate responsibility, moving away from a focus solely on shareholders to a broader consideration of all stakeholders. This shift is reshaping how companies, particularly those publicly traded on stock markets, approach and respond to pressures to improve their sustainability practices and disclosure. From a theoretical perspective, companies stand to gain in economic performance and address information imbalances by investing in environmental, social, and governance (ESG) activities over the long term. However, these benefits depend on the company's ability to manage the initial costs and navigate industry-specific dynamics associated with sustainability initiatives.

The COVID-19 pandemic has emphasized the importance of health care companies, and their contribution towards sustainable development goals. The health care industry can play a pivotal role in promoting health and well-being, reducing inequalities, and ensuring access to quality healthcare services for all. Nevertheless, this industry faces major sustainability challenges. Health care facilities significantly contribute to environmental pollution, generating large amounts of waste, consuming energy and water resources, and emitting greenhouse gases. The COVID-19 pandemic has also acted as a catalyst for the rapid digital transformation of various sectors, with the health care industry experiencing a particularly swift uptake of new digital services and products, as well as abnormal financial performance. This surge has led to the emergence of numerous online platforms tailored to meet evolving patient needs. Digital health technologies are often heralded as potential solutions to mitigate the environmental impact of health systems while maintaining quality and financial sustainability. Nevertheless, the extent to which the adoption of these technologies can effectively address greenhouse gas emissions, waste, and other sustainability challenges remains uncertain. Indeed, while digital technologies have advanced medical research and streamlined the health care supply chain, a crucial question persists on whether these advancements have been designed to effectively narrow or instead exacerbate the existing digital divide and health equity gap across the socioeconomic spectrum.

Regulations and norms have shifted towards requiring companies, especially listed corporations, to invest in sustainability and improve their profile and disclosure. Several reporting frameworks exist which are under development in a fast-moving environment which is directed towards improving transparency and information mechanisms between companies and all stakeholders. Improvement in sustainability can be measured quantitatively via the use of ESG ratings and scores that are provided by a handful of large providers, mostly from the USA and the UK.

From a management perspective, understanding whether improvements in ESG profile result in superior corporate performance is important. Yet, to date, limited evidence exists for the health care industry. Whether the company's economic performance and informative gaps (i.e., information asymmetry) between the company and the stakeholders will improve remains unclear. This book aims to address these propositions.

The first chapter addresses the social and normative context of sustainability, presenting the evolution of this concept first from a policy and then from a corporate perspective. It then proceeds with briefly introducing the financing mechanisms of sustainability and the implications that it has for corporate performance.

The second chapter delineates the role the health care sector plays for achieving a sustainable future, starting with its footprint and initiatives aimed at addressing its unique challenges. A section is subsequently dedicated to digital health, where its potential for contributing to sustainable goals and drawbacks are discussed in the context of the relevant literature.

The third chapter delves into how sustainability disclosure has evolved over time, what reporting frameworks and standards exist which can help companies meet imposed reporting requirements and improve their ESG profiles. The text then includes a discussion of the key ESG providers, the industry, and divergence between ESG rating and scoring systems.

Subsequently, the fourth chapter provides an evolutionary analysis of the sustainability and corporate performance of health care companies listed in European stock exchanges. Starting with a description of the structural characteristics of the companies included in the sample, a comparative evaluation of profitability, market values, and ESG profiles of these companies is presented and discussed before and after the European Directive 2014/95.

The following chapter focuses on the relationship between ESG score and economic performance. A review of key theories underlying that relationship is first presented, with two main conceptual frameworks being discussed: agency theory and stakeholder theory. Considering these theories and bases of evidence, a quantitative empirical contribution is subsequently presented using alternative performance evaluation measures.

The sixth chapter focuses on the concept of information asymmetry, proxied using the forecast error in final returns predictions by financial analysts. The chapter first discusses the signaling and the value relevance theory as conceptual frameworks for understanding how changes in ESG profile may influence information asymmetry. A statistical analysis conducted on the selected sample of European health care companies follows testing the respective hypothesis.

Finally, some concluding remarks are presented. These are based on the literature discussed and the reported study findings, which should all facilitate our understanding of the management implications of improvements in ESG profile by corporations and their effect on corporate performance.

# Contents

# List of Abbreviations

| | |
|---|---|
| CSR | Corporate Social Responsibility |
| CSRD | Corporate Sustainability Reporting Directive |
| EBITM | Earnings Before Interest and Tax Margin |
| EFRAG | European Financial Reporting Advisory Group |
| EPS | Earning Per Share |
| ESG | Environmental Social Governance |
| ESRS | European Sustainability Reporting Standards |
| EU | European Union |
| FASB | Financial Accounting Standards Board |
| FE | Forecast Error |
| GHG | Greenhouse Gas |
| GRI | Global Reporting Initiative |
| IASB | International Accounting Standards Board |
| IIRC | International Integrated Reporting Council |
| NHS | National Health Service |
| ROA | Return on Assets |
| ROE | Return on Equity |
| SASB | Sustainability Accounting Standards Board |
| SDG | Sustainable Development Goal |
| UK | United Kingdom |
| US | United States |

# Chapter 1
# Social and Normative Context of Sustainability

## Evolution of Sustainability

Since the 1960s, the model of endless economic growth has come under scrutiny and criticism, driven by a new awareness of the Earth's finite and limited capacity to provide resources (e.g., raw materials, water, energy) and absorb pollutants (e.g., solid waste, atmospheric emissions, wastewater). Among other initiatives, The Club of Rome established in April 1968 by an Italian entrepreneur Aurelio Peccei brought together scientists, economists, businesspeople, civil rights activists, senior international public officials, and heads of state, laying the groundwork for a new approach and a full awareness of the limits to economic growth. The Limits to Growth report (Meadows et al., 1972), and later the concept of *sustainable development,* emerged from this initiative delineating a new principle of economic and ecological stability. This principle was famously echoed by British economist Kenneth Ewart Boulding, who wittily remarked: "*Anyone who believes exponential growth can go on forever in a finite world is either a madman or an economist,*" highlighting a scathing critique of the existing model based on unlimited growth (Boulding, 1966). The central premise is that natural resources are finite, thus envisioning a system dedicated to endless growth is untenable, and improving living conditions must be achieved without increasing consumption. Sustainability thus suggests organizing collectively so that a decrease in goods production does not equate to a reduction in civilization levels.

Amidst growing antiwar sentiment and heightened societal awareness, a lack of trust in the ability of businesses to meet public needs and desires and an overall low confidence in the business sector reached a critical juncture in 1969 when a significant oil spill off the coast of Santa Barbara, California, triggered widespread protests throughout the United States, culminating in the inception of the inaugural Earth Day in 1970. This led to a revamped regulatory framework that would shape



P. Candio, *Sustainability and Corporate Performance in Health Care*, CSR, Sustainability, Ethics & Governance,
https://doi.org/10.1007/978-3-031-63985-2_1

corporate conduct and impose additional responsibilities for corporations (Latapí Agudelo et al., 2019).

The concept of sustainability was first officially discussed at the 1972 United Nations Conference on the Human Environment in Stockholm, where possible solutions for protecting the natural ecosystem were deliberated. During the conference an action plan and a Declaration were drafted, addressing humanity's rights and responsibilities regarding the environment and the need to ensure "compatible development" that maintains social well-being while safeguarding and distributing natural resources considering the needs of future generations (UN General Assembly, 1972).

In the 1980s, environmental disasters, particularly events such as Chernobyl and Exxon Valdez, prompted international agreements such as the Montreal Protocol in 1987, where it was agreed that the stratospheric ozone layer ought to be protected by phasing out production and consumption of ozone-depleting substances (Laine et al., 2021). In the same year, the World Commission on Environment and Development emphasized the cardinal concept of *sustainable development* in the landmark report "Our Common Future," also known as the Brundtland Report (Keeble, 1988). This report defined sustainable development as meeting the needs of the current generation without compromising the ability of future generations to meet their own needs, concept which became crucial in directing attention to sustainable development in legislative contexts and shaping new visions for it in the following decades.

By contrast, in the same time period, the Reagan and Thatcher administrations, in the USA and the UK, respectively, focused on alleviating pressure on corporations and addressing high levels of inflation (Feldstein, 2013). As governments scaled back regulations on corporate conduct, managers found themselves accountable in the public's eye to various interest groups. This shift prompted scholars to explore business ethics and the practical implementation of CSR to address the concerns of stakeholders such as shareholders, employees, and consumers, popularizing the concept of stakeholders (Carroll, 2008).

Over the following decade, significant global events reshaped the international perspective on social responsibility and sustainable development. These include the establishment of the European Environment Agency, the adoption of UN Agenda 21, and the United Nations Framework Convention on Climate Change in 1992, as well as the ratification of the Kyoto Protocol in 1997. These initiatives, including the creation of international organizations and the implementation of global agreements, reflected concerted international efforts to raise standards concerning climate-related issues and indirectly influence corporate conduct (Union of Concerned Scientists, 2018). For Europe, in 1992, the European Union adopted the Fifth Environmental Action Programme towards sustainability, addressing environmental challenges within the EU. Significant events followed, including the International Conference on Population and Development in 1994 and the Fourth World Conference on Women in 1995, focusing on gender equality and women's rights. In 1997, the UN General Assembly's Special Session, known as Rio + 5, assessed Agenda 21's implementation and identified new solutions for sustainable development, laying out the UN's role.

It is in the years 2000s' when the concept of sustainability, and particularly sustainable development and CSR gained formal international recognition and prominence. The adoption of the Millennium Declaration by the United Nations, along with its eight Millennium Development Goals (United Nations, 2023), marked a pivotal moment in setting the international agenda. Whereas the discourse surrounding these goals were not directly tied to CSR, the United Nations Development Program Identified CSR as a framework for fostering collaboration between the UN and the private sector, thereby enhancing the global recognition of the concept. The advocacy for CSR as a distinct European strategy emerged quickly after, with the presentation of a Green Paper by the European Commission titled "Promoting a European framework for Corporate Social Responsibility" (European Commission, 2001), underlying what CSR would entail in terms of responsibility of enterprises for their impacts on society and a company should do to meet that responsibility (European Commission, 2011).

The year 2015 marked a significant turning point with the adoption of the Paris Agreement (United Nations Framework Convention on Climate Change, n.d.) and the launch of the 2030 Agenda for Sustainable Development, accompanied by the adoption of seventeen Sustainable Development Goals (SDGs; United Nations, 2015). The seventeen SDGs encompass a broad spectrum of targets aimed at addressing global challenges and fostering sustainable development worldwide. However, the SDGs do not impose direct obligations on the private sector, rather they prompt countries to develop specific policies and regulations to indirectly pressure firms to adopt new business practices or enhance existing ones.

## Corporate's Responsibility: A Paradigm Shift

Since Adam Smith's contribution (Smith, 1776), a robust academic debate has revolved around the purpose and role of businesses within the economic and social system (Smith, 2003). However, the interplay between global economic development and social challenges emerged over the last two decades has led to significant shifts in society's expectations regarding the role and responsibilities of businesses, in particular corporations. Broadly speaking, two opposing groups of theories have battled it out for supremacy: the neoclassical (Solo, 1975) and the institutional theories (Hamilton, 1919). The former group tends to depict the firm as a constituent of the economic system merely driven by self-interest, where its primary aim is to maximize value for the owners, namely the shareholder. According to this philosophy, societal benefit is optimized through the most efficient allocation of scarce resources and based on the orthodox paradigm of neoclassical theory. There exists no institutional mechanism other than market allocation, both for input and output markets (Friedman, 1970). In essence, the sole recognized institution is the market, where prices are determined, while other institutions are deemed neither useful nor efficient, and may even hinder corporate performance. The most prominent figure from group is arguably Milton Friedman (shareholder theory; Friedman, 1962), though other major exponents include Alfred Marshall (theory of value; Frisch, 1950), Arthur Cecil Pigou (Economics of Welfare; Pigou, 1932), and many more.

The fundamental assumptions underlying the neoclassical doctrine of utility are threefold: exogeneity of preferences, irrelevance of motivations, and perfect market competition. This paradigm makes economics a closed science that requires no supplementation from external sources, whereby markets are places where rational and selfish individuals exchange goods and services to satisfy solely subjective preferences. All economic entities have exclusively material ends, and act for instrumental motivations, being completely devoid of an ethical, emotional, social, or altruistic dimension. The position of the neoclassical theory of the firm on CSR can be summarized by a statement made by Milton Friedman in 1970: "*There is one and only one social responsibility of business—to use its resources and engage in activities designed to increase its profits so long as it stays within the rules of the game which is to say, engages in open and free competition without deception or fraud*" (Friedman, 1970). The neoclassical philosophy justifies and describes the adoption of CSR behaviors only when their cost is offset by a clear benefit in terms of profits (Carson, 1993). While it is true that responsible business behavior depends on incentives that translate into financial results, it is also true that CSR acts are instrumental; they are only *means and not ends*, rather than embodied and experienced beliefs within companies.

According to the (neo) institutional group of theories, and in opposition to what is posited by the neoclassical theories, the market is embedded within other social institutions, and as such, it is promoted, created, and even shaped by social forces and institutions (Polanyi, 1944). The institutions create the rules by which firms and consumers, respectively, satisfy, rather than maximize, their profit and utility. This approach posits that institutions are not necessarily created to be socially and economically efficient (in Paretian sense; Montesano, 1997); rather, they are created to serve and preserve the interests of certain social groups and to establish new institutions. Institutions dictate all economic relationships of individuals regardless of processes of rationality and maximization. Within this group of theories, the earlier contributions by Berle and Means (1932), Coase (1937), and more recently Teubner (1988) and most notably Freeman's stakeholder theory (1984) stand out. It is also important to mention the seminal work of scholar Howard R. Bowen (1953) who arguably marked the beginning of modern studies on the topic of CSR. Bowen's perspective is based on the belief that businesses are vital centers of power and decision-making, and their actions influence citizens in various aspects of their lives.

The institutionalist perspective views the firm as an institution tasked with contributing to economic and social development. The firm is not seen as the exercise of a right, but rather as the fulfillment of a function to be carried out in the interest of those who establish the firm and the community in which it operates. The rationale for this conception revolves around the firm's power to impact the interests of the collective and the need to legitimize this power, responding to how it is used. In this perspective, profit appears to be just one of the firm's objectives. The ultimate goal of the institution in question is therefore the achievement of the common good, collective utility rather than solely individual gain. A framework developed by Carroll (1979, 1991) sought to reconcile the existence of economic, legal, social, and ethical objectives. Indeed, Carroll identifies four distinct responsibilities that a business must fulfill to survive, namely: economic, legal, ethical, and discretionary or philanthropic responsibilities. With Freeman's stakeholder theory, a crucial bridge towards a new conception of the firm was developed. Understood as an open and highly interconnected system between the business and stakeholders, the business lies among stakeholders, where interests attributable to various groups of individuals converge. The fundamental concepts underlying this theory are: the principle of interdependence (between business action and ethical action); relevance of values and the maximization of well-being (understood not only in economic terms) of all stakeholders.

Over the last two decades, other significant academic contributions have been made, such as that by Smith (2001) who highlighted the transformative landscape broadening the scope of social responsibility from a business standpoint to encompassing a wider array of stakeholders (Stiglitz, 2010). Marrewijk (2003) depicted this new approach as a strategic response to evolving corporate challenges which stem from the changing roles and responsibilities of each sector of society. Chandler and Werther (2013) further delved into this concept, focusing on the implementation of strategic CSR as part of brand management to uphold legitimacy in a globalized brand landscape. Building on this premise, Porter and Kramer (2006) posited that companies can gain a competitive edge through strategic CSR by creating shared value for society while enhancing their competitiveness. They advocated for a strategic approach wherein companies first assess the social impact of their value chain internally and externally, enabling them to adapt their business strategies accordingly. Porter and Kramer (2011) subsequently argued that the imperative for creating shared value stems from conventional business strategies, which often overlook the broader factors crucial for long-term success.

## Sustainable Finance

Achieving a sustainable future requires collective commitment and significant financial investments. In this respect, sustainable finance has emerged as a means to foster a more sustainable and inclusive global economy by aligning financial activities with long-term sustainable development goals while minimizing negative impacts on the planet and society (Roncalli, 2024). Sustainable finance plays a pivotal role in incentivizing and rewarding companies committed to ESG principles. By integrating sustainability criteria into investment decisions, sustainable finance channels capital towards companies with responsible ESG practices, thus fostering a transition to a more sustainable economy.

The European Commission defines sustainable finance as *"Sustainable finance refers to the process of taking environmental, social and governance considerations into account when making investment decisions in the financial sector, leading to more long-term investments in sustainable economic activities and projects. Environmental considerations might include climate change mitigation and adaptation, as well as the environment more broadly, for instance the preservation of biodiversity, pollution prevention and the circular economy. Social considerations could refer to issues of inequality, inclusiveness, labor relations, investment in people and their skills and communities, as well as human rights issues. The governance of public and private institutions—including management structures, employee relations and executive remuneration—plays a fundamental role in ensuring the inclusion of social and environmental considerations in the decision-making process."* (European Commission, n.d.).

In brief, the goal of sustainable finance is to redirect capital flows towards investments that generate positive social and environmental outcomes alongside *financial returns*. In addition, sustainable finance seeks to promote financial stability and resilience in the face of environmental and social challenges. By integrating sustainability considerations into financial decision-making processes, sustainable finance helps identify and mitigate emerging *risks*, such as climate-related disruptions, natural disasters, and social unrest. This proactive approach to risk management can enhance the resilience of financial institutions, investors, and the broader economy.

Sustainable finance is crucially based on a long-term investment perspective (Angelini, 2022), encompassing the range of investment choices that ought to be considered, not solely based on profit criteria or in line with the investor's risk profile, but also taking into account corporations' impacts on the real economy, as well as on society and the environment. In particular, events related to climate change and its adaptation, biodiversity and natural ecosystems, as well as the development of human capital and addressing of inequalities. Balancing environmental, social, and economic factors would make it possible to intercept negative externalities which can translate into costs or opportunities for companies, investors, and other interested parties.

Sustainable finance is part of a wide array of initiatives including ESG Investing, Green Finance, Climate Risk, Socially Responsible Investing, Responsible Investment, and Sustainable Development (Roncalli, 2024), all aiming towards the realization of sustainable development. According to the United Nations Environment Programme (2016), terms such as "green finance," "sustainable finance," "climate finance," and "low-carbon finance" and many more relate to an overlapping territory of issues, applied to financial decision-making and flows, encompassing environmental, social, economic, and governance considerations.

Climate finance is commonly associated with the United Nations Framework Convention on Climate Change, which defines it as "local, national or transnational financing, which may be drawn from public, private and alternative sources of financing. Climate finance is critical to both reduce emissions and to allow countries to adapt to the adverse effects and reduce the impacts of climate change." Green finance, on the other hand, typically encompasses a broader scope than climate finance, addressing additional environmental objectives and risks. It often involves a stronger emphasis on greening diverse streams of private investment rather than primarily focusing on public and publicly leveraged financial flows.

## Implications for Corporate Performance

### *Economic Performance*

The economic performance of a company refers to its ability to generate profit and create value for stakeholders within a given period. It encompasses various financial metrics that assess the company's efficiency, profitability, and economic stability, extending beyond financial performance metrics to encompass broader aspects such as productivity, innovation, market value, and strategic management.

The increasing pressure for sustainable business practices has heightened consciousness among stakeholders, translating into greater consumer awareness (Boufounou et al., 2023), evolving investor expectations (Kräussl et al., 2023), and stringent regulatory mandates (European Union, 2022). This pressure is compelling companies, especially large corporations which source funds from capital markets, to reevaluate their operational paradigms and embark on a trajectory towards sustainability, integrating ESG considerations into their core strategies (de Souza Barbosa et al., 2023). Although with marked heterogeneity across industries (Wang et al., 2023), these measures encompass a broad spectrum of actions, including the adoption of eco-friendly practices throughout their value chains, active engagement with stakeholders to address concerns and incorporate feedback, investment in innovation to develop sustainable solutions, and transparent reporting of ESG performance. The rationale behind this concerted effort lies in the recognition that investing in long-term sustainability may yield benefits for companies in the long term, particularly in terms of economic performance.

Although investing in ESG-related initiatives may hold the promise to offer long-term benefits, including resilience to future shocks (Yahya, 2023), it is important to recognize that such endeavors often entail significant upfront costs and uncertainties. For instance, transitioning to renewable energy sources, enhancing supply chain transparency, or implementing sustainable production processes can involve infrastructure investments in infrastructure and capital-intensive upgrades. Moreover, there are operational expenses associated with maintaining sustainable practices, such as monitoring and reporting ESG metrics, conducting audits, and compliance with evolving regulations. These expenditures can strain budgets, posing a barrier to entry into sustainable practices (Liou et al., 2023).

Additionally, investing in ESG may require organization-level changes and investments in activities such as employee training, stakeholder engagement, and management initiatives and resistance to these changes or skepticism from shareholders may further exacerbate these challenges (Choi et al., 2020). Moreover, there are concrete risks of double-edged dynamics associated with ESG investments. These include drawbacks such as increased market volatility, regulatory uncertainty, and reputational risks if sustainability goals are not met or perceived as greenwashing (Acuti et al., 2022).

While from a theoretical standpoint, investing in sustainability may improve the company's economic performance, at least in the long run, whether this will manifest in reality remains uncertain (Gupta & Das, 2022). In addition, the approach to sustainability and effects of improvements in ESG profile will differ across industries (Galant & Cadez, 2017), reflecting the unique environmental impacts, social responsibilities, and governance structures inherent to each sector. Industries with high environmental footprints, such as manufacturing or energy production, face greater pressure to adopt sustainable practices to mitigate negative impacts on ecosystems and communities and therefore activities directed towards addressing those issues may be rewarded to a greater degree than other ESG-relevant dimensions. In contrast, service-based industries may prioritize social sustainability through initiatives promoting diversity, inclusion, and corporate social responsibility. Finally, depending on transaction costs and contextual market competition, some industries may find it easier to implement sustainable practices in certain domains due to technological innovations available in the market, whereas others may encounter barriers such as cost constraints or lack of incentives (Alayón et al., 2022). Chapter 5 addresses this question for the Health Care industry.

### *Information Asymmetry*

Changes to sustainability profiles may have effects on mediatory factors on the pathway to economic performance, in particular information mechanisms. Specifically, improved ESG profile and associated disclosure will 'signal' the respective changes to information intermediaries such as financial analysts, rating agencies, and company data providers (Wen et al., 2022), who in turn aggregate, analyze, and disseminate data for investors and financial intermediaries. Improved disclosure will consequently reduce the so-called *information asymmetry* existing between the company and the market and stakeholders (Al Natour et al., 2022).

From a theoretical standpoint, reduced information asymmetry will reflect better information available to financial analysts (Talmor, 1981), which will enhance their ability to generate accurate predictions regarding the company's future economic performance. This will in turn enhance the corporation's credibility (Pope & Wang, 2023) and transparency with investors and other stakeholders: by consistently meeting forecasted earnings and performance targets, the corporation will build trust and confidence among shareholders and improve its credibility fostering stronger investor relations, reducing financing costs, and enhancing access to capital markets, all of which potentially facilitating future growth and expansion initiatives (Zhao et al., 2023).

However, whether reduced information asymmetry will derive from improvement in a company's ESG profile in practice remains unclear to date. Similar to what was discussed for economic performance, regulatory constraints aside, industry-specific effects will also characterize these dynamics, inevitably increasing decision uncertainty regarding the choice of investing in sustainability (Rossi & Candio, 2023). Chapter 6 addresses this question for the Health Care industry.

# References


Acuti, D., Pizzetti, M., & Dolnicar, S. (2022). When sustainability backfires: A review on the unintended negative side-effects of product and service sustainability on consumer behavior. *Psychology & Marketing, 39*, 1933–1945. https://doi.org/10.1002/mar.21709

Al Natour, A. R., Meqbel, R., Kayed, S., & Zaidan, H. (2022). The role of sustainability reporting in reducing information asymmetry: The case of family- and non-family-controlled firms. *Sustainability, 14*, 6644. https://doi.org/10.3390/su14116644

Alayón, C. L., Säfsten, K., & Johansson, G. (2022). Barriers and enablers for the adoption of sustainable manufacturing by manufacturing SMEs. *Sustainability, 14*, 2364. https://doi.org/10.3390/su14042364

Angelini, P. (2022). *Long-term investing and sustainable finance: Challenges and perspectives. Bank of Italy*. Accessed March 10, 2024, from https://www.bancaditalia.it/pubblicazioni/interventi-direttorio/int-dir-2022/Angelini-11-luglio-2022.pdf

Berle, A. A. Jr., & Means, G. C. (1932). *The modern corporation and private property* (Macmillan 1933).

Boufounou, P., Moustairas, I., Toudas, K., et al. (2023). ESGs and customer choice: Some empirical evidence. *Circular Economy and Sustainability, 3*, 1841–1874. https://doi.org/10.1007/s43615-023-00251-8

Boulding, K. (1966). The economics of the coming spaceship earth. In *Environmental quality in a growing economy* (pp. xx–xx). Johns Hopkins Press.

Bowen, H. R. (1953). *Social responsibilities of the businessman*. University of Michigan.

Carroll, A. B. (1979). A three-dimensional model of corporate social performance. *Academy of Management Review, 4*(4), 497–505.

Carroll, A. B. (1991). The pyramid of corporate social responsibility: Toward the moral management of organizational stakeholders. *Business Horizons, 34*(4), 39–48.

Carroll, A. B. (2008). A history of corporate social responsibility: Concepts and practices. In A. M. Andrew Crane, D. Matten, J. Moon, & D. Siegel (Eds.), *The Oxford handbook of corporate social responsibility* (pp. 19–46). Oxford University Press.

Carson, T. (1993). Friedman's theory of corporate social responsibility. *Business & Professional Ethics Journal, 12*(1), 3–32. http://www.jstor.org/stable/27800897

Chandler, D., & Werther, W. B. (2013). *Strategic corporate social responsibility: Stakeholders, globalization, and sustainable value creation* (3rd ed.). Sage.

Choi, D., Chung, C. Y., Seyha, T., & Young, J. (2020). Factors affecting organizations' resistance to the adoption of blockchain technology in supply networks. *Sustainability, 12*, 8882. https://doi.org/10.3390/su12218882

Coase, R. (1937). The nature of the firm. *Economica, 4*, 386–405.

de Souza Barbosa, A., da Silva, M. C. B. C., da Silva, L. B., et al. (2023). Integration of Environmental, Social, and Governance (ESG) criteria: Their impacts on corporate sustainability performance. *Humanities and Social Sciences Communications, 10*, 410. https://doi.org/10.1057/s41599-023-01919-0

European Commission. (2001). *Promoting a European framework for Corporate Social Responsibility*. Accessed March 07, 2024, from https://ec.europa.eu/commission/presscorner/api/files/document/print/en/doc_01_9/DOC_01_9_EN.pdf

European Commission. (2011). *Corporate social responsibility: A new definition, a new agenda for action*. (MEMO/11/732, MEMO/11/734 and MEMO/11/735). European Commission. Accessed March 07, 2024, from http://europa.eu/rapid/press-release_MEMO-11-730_en.htm

European Commission. (n.d.). *Overview of sustainable finance*. Accessed March 10, 2024, from https://finance.ec.europa.eu/sustainable-finance/overview-sustainable-finance_en

European Union. (2022). *Directive (EU) 2022/2464 of the European Parliament and of the Council of 14 December 2022 amending Regulation (EU) No 537/2014, Directive 2004/109/EC, Directive 2006/43/EC and Directive 2013/34/EU, as regards corporate sustainabil-*

*ity reporting*. Accessed March 07, 2024, from https://eur-lex.europa.eu/legal-content/EN/TXT/?uri=CELEX:32022L2464

Feldstein, M. (2013). *The Reagan-Thatcher revolution*. Accessed November 9, 2023, from https://www.dw.com/en/the-reagan-thatcher-revolution/a-16732731

Freeman, E. R. (1984). *Strategic management: A stakeholder approach*. Pitman.

Friedman, M. (1962). *Capitalism and freedom*. University of Chicago Press.

Friedman, M. (1970, September 13). *A Friedman doctrine: The social responsibility of business is to increase its profits*. The New York Times Magazine.

Frisch, R. (1950). Alfred Marshall's theory of value [Review of Principles of Economics, Book V (7th ed), by A. Marshall]. *The Quarterly Journal of Economics, 64*(4), 495–524. https://doi.org/10.2307/1884385

Galant, A., & Cadez, S. (2017). Corporate social responsibility and financial performance relationship: A review of measurement approaches. *Economic Research-Ekonomska Istraživanja, 30*(1), 676–693. https://doi.org/10.1080/1331677X.2017.1313122

Gupta, J., & Das, N. (2022). Multidimensional corporate social responsibility disclosure and financial performance: A meta-analytical review. *Corporate Social Responsibility and Environmental Management, 29*(4), 731–748. https://doi.org/10.1002/csr.2237

Hamilton, W. H. (1919). The institutional approach to economic theory. *The American Economic Review, 9*(1), 309–318. http://www.jstor.org/stable/1814009

Keeble, B. R. (1988). The Brundtland report: "Our Common Future". *Medicine and War, 4*(1), 17–25. Retrieved from http://www.jstor.org/stable/45353161

Kräussl, R., Oladiran, T., & Stefanova, D. (2023). A review on ESG investing: Investors' expectations, beliefs and perceptions. Journal of Economic Surveys, Special Issue. https://doi.org/10.1111/joes.12599

Laine, M., Tregidga, H., & Unerman, J. (2021). Sustainability reporting, history, frameworks and regulation. In M. Laine, H. Tregidga, & J. Unerman (Eds.), *Sustainability accounting and accountability* (pp. 83–102). Routledge.

Latapí Agudelo, M. A., Jóhannsdóttir, L., & Davídsdóttir, B. (2019). A literature review of the history and evolution of corporate social responsibility. *International Journal of Corporate Social Responsibility, 4*(1). https://doi.org/10.1186/s40991-018-0039-y

Liou, J. J. H., Liu, P. Y. L., & Huang, S.-W. (2023). Exploring the key barriers to ESG adoption in enterprises. *Systems and Soft Computing, 5*, 200066. https://doi.org/10.1016/j.sasc.2023.200066

Marrewijk, M. (2003). *Concepts and definitions of CSR and corporate sustainability: Between agency and communion* (Vol. 44).

Meadows, D. H., Meadows, D. L., Randers, J., & Behrens, W. W., III. (1972). *The limits to growth*. Universe Books.

Montesano, A. (1997). Pareto's analysis of efficiency and its interpretation. *History of Economic Ideas, 5*(3), 7–18. http://www.jstor.org/stable/23722572

Pigou, A. C. (1932). *The economics of welfare. Library of economics and liberty*. Accessed March 10, 2024, from http://www.econlib.org/library/NPDBooks/Pigou/pgEW.html

Polanyi, K. (1944). *The great transformation*. Rinehart.

Pope, P. F., & Wang, T. (2023). Analyst ability and research effort: Non-EPS forecast provision as a research quality signal. *Review of Accounting Studies, 28*, 1263–1315. https://doi.org/10.1007/s11142-023-09791-8

Porter, M. E., & Kramer, M. R. (2006, December). *Strategy & society* (pp. 1–16). Harvard Business Review.

Porter, M. E., & Kramer, M. R. (2011, January–February). *Creating shared value*. Harvard Business Review.

Roncalli, T. (2024). *Handbook of sustainable finance*. Working papers hal-04370824. Accessed March 10, 2024, from https://ideas.repec.org/p/hal/wpaper/hal-04370824.html

Rossi, P., & Candio, P. (2023). The independent and moderating role of choice of non-financial reporting format on forecast accuracy and ESG disclosure. *Journal of Environmental Management, 345*, 118891. https://doi.org/10.1016/j.jenvman.2023.118891

Smith, A. (1776). *An inquiry into the nature and causes of the wealth of nations* (History of Economic Thought Books No. smith1776). McMaster University Archive for the History of Economic Thought.

Smith, N. C. (2001). Changes in corporate practices in response to public interest advocacy and actions. In P. N. B. a. G. T. Gundlach (Ed.), *Handbook of marketing and society*. Thousand Oaks.

Smith, H. J. (2003, July 15). *The Shareholders vs. Stakeholders Debate*. MIT Sloan Management Review, (Summer 2003).

Solo, R. A. (1975). Neoclassical economics in perspective. *Journal of Economic Issues, 9*(4), 627–644. http://www.jstor.org/stable/4224456

Stiglitz, J. (2010). *Freefall: America, free markets, and the sinking of the world economy*. W. W. Norton & Company.

Talmor, E. (1981). Asymmetric information, signaling, and optimal corporate financial decisions. *The Journal of Financial and Quantitative Analysis, 16*(4), 413–435. https://doi.org/10.2307/2330363

Teubner, G. (1988). Enterprise corporativism: New industrial policy and the essence of the legal person. *American Journal of Corporate Law, 36*, 130–155.

UN General Assembly. (1972, December 15). *United Nations Conference on the Human Environment, A/RES/2994*. Accessed March 10, 2024, from https://www.refworld.org/legal/resolution/unga/1972/en/9934

Union of Concerned Scientists. (2018). *The IPCC: Who are they and why do their climate reports matter?* https://ucsusa.org/global-warming/science-and-impacts/science/ipcc-backgrounder.html

United Nations. (2015). *The sustainable development agenda*. Accessed March 07, 2024, from https://www.un.org/sustainabledevelopment/development-agenda/

United Nations. (2023). *Millennium development goals*. Accessed March 07, 2024, from https://unis.unvienna.org/unis/topics/related/2015/millennium-development-goals-mdgs.html

United Nations Environment Programme. (2016). *Design of a sustainable financial system: Definitions and concept*. Accessed March 07, 2024, from https://wedocs.unep.org/bitstream/handle/20.500.11822/10603/definitions_concept.pdf?sequence=1&amp%3BisAllowed=

United Nations Framework Convention on Climate Change. (n.d.). *The Paris Agreement*. Accessed March 07, 2024, from https://unfccc.int/process-and-meetings/the-paris-agreement

Wang, Y., Lin, Y., Fu, X., & Chen, S. (2023). Institutional ownership heterogeneity and ESG performance: Evidence from China. *Finance Research Letters, 51*, 103448. https://doi.org/10.1016/j.frl.2022.103448

Wen, H., Ho, K. C., Gao, J., & Yu, L. (2022). The fundamental effects of ESG disclosure quality in boosting the growth of ESG investing. *Journal of International Financial Markets, Institutions and Money, 81*, 101655. https://doi.org/10.1016/j.intfin.2022.101655

Yahya, H. (2023). The role of ESG performance in firms' resilience during the COVID-19 pandemic: Evidence from Nordic firms. *Global Finance Journal, 58*, 100905. https://doi.org/10.1016/j.gfj.2023.100905

Zhao, Y., Yao, Z., Li, Y., & Zhou, P. (2023). Can high-quality interactions lower the cost of debt? Insights from interactive investor platforms. *Finance Research Letters, 58*(Part C), 104600. https://doi.org/10.1016/j.frl.2023.104600

# Chapter 2
# Sustainability in Health Care

## Background

The distribution of private and public health care systems varies considerably across the globe and is shaped by governmental regulations, societal norms, and economic circumstances. On the one hand, health care can be administered and funded by the public sector, with government-backed health care frameworks catering to a substantial portion of or all population needs, such as in the case of the National Health Service in the United Kingdom, Canada's publicly subsidized health care system, and Australia's Medicare program. In these settings, medical services are provided 'free at the point of need' or at reduced cost, with financial support originating from taxes or social insurance schemes. Conversely, in other countries, the private sector plays a major or an exclusive role, where health care facilities are operated by privately held hospitals, clinics, and health care practitioners and providers. In these settings, patients may incur direct expenses for health services or hold private health insurance plans, which may be procured independently or supplied by employers. The United States stands as the most prominent example of such setting, where private health insurance significantly underwrites health care expenditures.

From a supply side argument, the health care industry can be broadly classified into the following macro sectors: pharmaceuticals, biotechnology, medical equipment, distribution, facilities, and managed health care. At a macro level, this industry has exhibited both stable and reactive characteristics within the global economy over the last decades. Essential products like medications, therapies, and medical facilities are indispensable for patient health, contributing to the sector's secular growth as innovative treatments address a wider range of illnesses. However, this growth potential is tempered by increased volatility, driven by the industry's susceptibility to shifts in government policies. In the USA, for example, Biotech and pharmaceutical companies face heightened sensitivity due to their reliance on insurance coverage and reimbursement for product sales (Blackstone & Fuhr, 2007).



P. Candio, *Sustainability and Corporate Performance in Health Care*, CSR, Sustainability, Ethics & Governance,
https://doi.org/10.1007/978-3-031-63985-2_2

The COVID-19 pandemic profoundly impacted the health care industry worldwide. It led to a surge in demand for health care services, straining resources and prompting a shift in priorities towards treating COVID-19 patients. This resulted in a surge in the adoption of telehealth and remote services, disrupting routine care delivery but accelerating innovation in digital health. The pandemic reshaped this industry by emphasizing the need for adaptability, innovation, and collaboration to address future health crises effectively, while at the same time contributing to the achievement of sustainable goals. Indeed, the health care industry plays a crucial role in achieving the UN Sustainable Development Goals (SDGs). By addressing health care provision, the industry can contribute to good health and well-being (SDG 3), gender equality (SDG 5), decent work and economic growth (SDG 8), and partnerships (SDG 17). On the path towards achieving a sustainable future, there is a contribution to be made by private health care companies in terms of improved transparency and commitment to sustainability.

## The Industry's Footprint

The health care industry makes a major contribution to the climate crisis, most predominantly in terms of its greenhouse gas (GHG) emissions and waste (Lenzen et al., 2020).

### *Greenhouse Gas Emissions*

Health care systems are responsible for around 5% of the total GHG emissions worldwide, with country-level estimates suggesting a share range between 3.3% and 8.1% (Pichler et al., 2019). Although varying greatly in magnitude across countries, health care systems emit GHGs both directly and indirectly during the provision of care and the acquisition of goods, services, and technologies from carbon-intensive supply chains. In particular, the industry contributes to total GHG emissions through its energy usage, transportation activities, and the production, utilization, and disposal of products.

A common framework- for the allocation and reporting of GHG emissions is the Greenhouse Gas Protocol (GHG Protocol, 2024). This protocol lists the following classes of GHGs to be included in footprint calculations, ordered in decreasing level of contribution to global warming, as follows: carbon dioxide, methane, nitrous oxide, hydrofluorocarbons, perfluorocarbons, and sulfur hexafluoride. Based on this accounting Protocol, GHG emissions can be grouped according to the following scope categories:

- Scope 1: Direct emissions stemming from activities under the organization's control. This could involve on-site fuel combustion, manufacturing and process emissions, refrigerant losses, or company vehicles.
- Scope 2: Indirect emissions from the consumption of electricity, heat, or steam that you purchase and utilize. Although not directly controlled, your use of energy contributes indirectly to $CO_2$ emissions.
- Scope 3: All other indirect emissions from sources beyond your direct control. This category encompasses emissions associated with activities throughout the value chain for which the organization bears indirect responsibility. This includes the procurement of goods and services, usage of sold goods, business travel, commuting, waste disposal, and water consumption.

Mapping health care providers and spending categories to the three GHG scope levels, a recent report (Health Care Without Harm, 2019) presented a framework to facilitate interpretation. Several key findings were identified from that study, which overall supported the following conclusion: "*if health care were a country, it would be the fifth largest emitter on the planet*." The key findings were:

- Emissions originating directly from health care facilities—Scope 1 emissions—account for 17% of the sector's global impact. Another 12% is attributed to indirect emissions from purchased electricity, steam, cooling, and heating (Scope 2). The majority, at 71%, arises from Scope 3, mainly stemming from the health care supply chain, including the production, transportation, utilization, and disposal of goods and services consumed by the industry. Considering all three scopes, over half of the health sector's environmental impact is linked to energy usage, particularly the consumption of electricity, gas, steam, and air conditioning supply, along with operational emissions.
- Whereas China, the USA, and the European Union emerge as the leading contributors to the health care sector's climate footprint in absolute terms, when emissions are assessed on a per capita basis—a crucial metric for understanding and addressing climate change equitably—, the perspective shifts. Globally, the average emissions per capita for health care activities stand at 0.28 tCO2. the United States leads both in absolute emissions (546 Mt) and per capita emissions (1.72 metric tons), producing 57 times more emissions per person compared to India which ranks seventh in absolute terms (39 Mt. CO2 e) and exhibits the lowest health sector-related emissions per capita (0.03 metric tons). Other top health sector emitters such as Australia, Canada, and Switzerland emit between 30 to 50 times more, per capita, than India. China boasts per capita emissions at 0.25, slightly below the global average (0.28).
- 40% of the sector's climate footprint originates from electricity and thermal power supply associated with health care-related activities. This encompasses energy procured by facilities (Scope 2), as well as that obtained by manufacturers and suppliers of goods and services for the sector (Scope 3, e.g. pharmaceutical and medical device industries). Additionally, another 13% stems from the so-called health sector operational emissions, bringing the total to 53%.

– A notable correlation exists, though not linear, between a country's health sector's climate footprint and its health spending: the higher the expenditure (measured as a percentage of a country's GDP), the higher the per capita emissions. However, several other factors play significant roles, particularly the energy intensity of a country's economy and the emissions intensity of its energy system. These factors contribute to nuances in the relationship between health spending and emissions.

Against this backdrop, the industry has been pressured to act and some evidence on this effect has recently started to emerge. Case in point for England, research has showed that, in 2019, the health care sector emitted 25 megatons of carbon dioxide equivalent: this marked a 26% decrease since 1990, with emissions per inpatient finished admission episode also dropping by 64%. Breaking this figure down, 62% derived from the supply chain, 24% from direct care delivery, 10% from staff commute and patient/visitor travel, and 4% from private healthcare services commissioned by the National Health Service (Tennison et al., 2021). A recent study employing a content analytical approach to examine pharmaceutical company climate change targets, GHG emissions, and strategies to reduce them found that the analyzed companies showed largely favorable reductions in scope 1 (in-house) and scope 2 (purchased energy), with variable results in scope 3 (supply chain) emissions (Booth et al., 2023). Efforts to cut emissions involved streamlining manufacturing and distribution processes, as well as responsibly sourcing energy, water, and raw materials.

## *Waste*

The UN Basel Convention considers health care waste among the most dangerous types of waste (Basel Convention, 2020). Approximately 85% of the waste produced from health care activities is classified as regular, non-dangerous waste (e.g., paper and cardboard, packaging and food waste), including plastic. For the latter, Penga et al. (2021) forecasted that worldwide 193 countries would see a rise in plastic waste by 8.4 million tons due to COVID-19-related activities, marking a 10% increase from the pre-pandemic baseline since WHO declared the disease a global pandemic in March 2020. This surge in plastic waste during the pandemic was primarily attributed to health care institutions, accounting for approximately 87.4%, including personal protective equipment (e.g.; masks, gloves, and face shields), online shopping packaging materials, and virus test kits, contributing 7.6%, 4.7%, and 0.3%, respectively. With approximately 280 million confirmed COVID-19 cases by the end of 2021, the total volume of medical waste was estimated to be around 11 million tons, with about 34,000 tons potentially entering the sea (IHME, 2021).

The remaining 15% is deemed hazardous material, as being potentially infectious, toxic, or radioactive (WHO, 2018a). Hazardous waste includes a variety of products and by-products, namely sharps (e.g., syringes, needles, disposable scalpels and blades), infectious material (e.g., contaminated with blood and other bodily fluids, such as those from discarded diagnostic samples, cultures, and stocks of infectious agents from laboratory activities or waste from patients with infections, like swabs, bandages, and disposable medical devices), pathological waste (such as human tissues, organs, or fluids), pharmaceutical waste (e.g., expired, unused, and contaminated drugs and vaccines), cytotoxic waste (substances with genotoxic properties, that is, highly hazardous substances that are mutagenic, teratogenic, or carcinogenic), chemical waste (e.g., solvents and reagents utilized for laboratory preparations, disinfectants, sterilant, and heavy metals found in medical devices (for instance, mercury in broken thermometers) as well as batteries), and radioactive waste (i.e., contaminated by radionuclides including radioactive diagnostic material or radiotherapeutic material).

In recent years, there has been an increasing debate surrounding the use of incineration for health care waste because of the production of dioxins, furans, and co-planar polychlorobiphenyls during the process (Zikhathile et al., 2022). Dioxins, furans, and co-planar PCBs are toxic by-products of various industrial processes that are generated as emissions or fly ash during incineration, particularly at temperatures below 800 °C. While some types of dioxins and furans may pose minimal threats, these substances are persistent and resistant to degradation, potentially accumulating in the food chain. The global burden of mortality and morbidity from exposure to dioxins and furans remains challenging to estimate due to insufficient data on exposure and risk assessment. However, many countries have implemented strict emission standards for dioxins and furans, leading to a significant reduction in their release into the environment (Liu et al., 2023).

Inadequate waste management greatly increases the risk of serious harm. (Health Protection Agency, 2010; Rushton, 2003; WHO, 2015) To name a few of the potential risks and issues:

- Improperly constructed or maintained landfills pose a risk of contaminating drinking water sources.
- Disposal facilities that lack proper design, management, or maintenance may present occupational hazards.
- Incineration of waste containing chlorine can result in the release of pollutants such as dioxins and furans, known human carcinogens with harmful health effects.
- Incineration of materials containing heavy metals like lead, mercury, and cadmium can spread toxic metals in the environment, leading to bioaccumulation.
- In areas where animals graze near incinerators, there is a potential risk of metal ingestion by humans.
- Chemicals used in waste treatment processes represent potential sources of water pollution through the sewer system.
- Waste can serve as a potential reservoir for harmful organisms, posing risks of pathogen transmission and serious public health implications.

- Waste can cause injuries like radiation burns, sharps-related wounds, and poisoning from toxic elements, wastewater, and pharmaceutical products.
- Expired medications, if not properly disposed of, may be collected and resold, posing public health risks.
- Long-term, low-level exposure to dioxins and furans can impair the immune, endocrine, and reproductive systems, manifesting after prolonged exposure. Short-term, high-level exposure to these substances may lead to skin lesions and altered liver function.

WHO, in collaboration with the United Nations Development Programme, has been actively supporting countries in developing optimal health care waste management strategies (United Nations Development Programme, 2023). This assistance aims to safeguard public health and the global environment from the adverse impacts of hazardous and toxic wastes through several initiatives:

1. Establishing model facilities and programs to showcase the implementation of top-tier health care waste management practices.
2. Deploying and assessing commercially available, non-incineration waste treatment technologies tailored to the specific needs of each country.
3. Developing, testing, producing, and deploying cost-effective, small-scale non-incineration technologies for widespread use.
4. Introducing and evaluating the utilization of mercury-free devices in model facilities.
5. Instituting or improving training programs to enhance capacity for implementing best practices and technologies, both within designated model facilities and beyond.
6. Reviewing and revising pertinent policies to align with current best practices and advancements.
7. Disseminating project outcomes and optimal practices to stakeholders for wider dissemination and adoption.

## Local and International Initiatives

The way health care organizations interact with all stakeholders has an impact on sustainability and can contribute to the development of sustainability-oriented health care systems (Maghsoudi et al., 2020). Indeed, health care systems involve several economic entities, such as professionals (both clinical and non-clinical), managers, patients, product providers, scientists, and government officials. This underscores the importance of implementing collaborative models based on shared goals when crafting policies, strategies, and objectives, to ensure social sustainability (Macassa & Tomaselli, 2020). Collaboration among multiple stakeholders fosters greater credibility, commitment, accountability, and support, enhancing the legitimacy of the health care system (Mariani et al., 2022). Nevertheless, the health care sector is facing issues of low trust and limited cooperation, both among professionals and within organizations (Alderwick et al., 2021).

To tackle these challenges, collaborative initiatives have been implemented. An example of a new system-wide approach to the delivery of quality health care across the whole system is that reported by Amoo et al. (2016). These authors focused on the role that the Leeds Institute for Quality Healthcare (Leeds, UK)—a partnership between the University of Leeds, NHS Organisations across the city and Leeds City Council—is playing in facilitating advancements in quality care by empowering clinicians to develop shared expertise, alongside establishing a robust framework for professional accountability which involved utilizing data to scrutinize decision-making processes.

Among institutional-level international initiatives for cross-sector collaborations, the Sustainable Markets Initiative, initiated by His Majesty King Charles III during his time as Prince of Wales in 2020 at the World Economic Forum in Davos, is worthy of note (Sustainable Markets Initiative, 2024a). This initiative aims to serve as a global private sector entity dedicated to advancing sustainable practices, by leveraging its ability to gather leading organizations from various industries, financial services, and governments and therefore drive innovation, hasten progress, and ensure a fair, sustainable, and prosperous future. Unveiled in 2021 at the One Planet Summit (2023), the Terra Carta—a mandate setting ambitious and practical action to help the private sector accelerate their progress towards a sustainable future—embodies a set of guiding principles and ambitious targets through 2030, prioritizing the integration of Nature, People, and Planet in global value creation (Sustainable Markets Initiative, 2024b). To date, the Terra Carta has boasted over 500 CEO-level supporters and catalyzed the creation of the Terra Carta Design Lab which serves as a roadmap for fostering collaboration among the public, private, and philanthropic sectors.

Based on this framework, the Sustainable Markets Initiative Health Systems Task Force was established during COP26 with the primary objective of hastening the implementation of carbon-neutral, patient-focused healthcare systems that enhance individual, societal, and global well-being. This collaborative effort involves CEOs and leaders from various organizations, including AstraZeneca, GSK, Merck, Novo Nordisk, Roche, Samsung Biologics, Sanofi, the Karolinska Institutet, NHS England, the Sustainable Healthcare Coalition, UNICEF, the University of Pavia, and the WHO. Additionally, the Task Force partners with the WHO's Alliance on Transformative Action on Climate and Health (ATACH, WHO, 2024). The aim of this alliance is to bolster climate resilience and decrease emissions from health care systems, by a parternship built across over 60 countries pledging their commitment at the Minister of Health level. In preparation for COP28, members of the Task Force introduced pioneering commitments, initiatives, and suggestions aimed at achieving short-term objectives and facilitating the shift towards sustainable, carbon-neutral health care.

Another relevant international initiative is Health Care Without Harm. In collaboration with a multinational professional services firm (Arup, 2024), this initiative recently formed a technical advisory panel tasked with overseeing the creation of the framework, methodology, analysis, and additional research for a global road map for health care decarbonization (Health Care Without Harm, 2021a). As a navigational tool for achieving zero emissions with climate resilience and health equity, a road map was proposed with the aim to ensure precision and reliability, incorporate healthcare-specific elements into climate footprint assessments, adhere to industry standards, accommodate variations in regional healthcare systems, and ensure acceptance by relevant stakeholders.

According to projections presented within a report by this initiative (Health Care Without Harm, 2021b), under a "business as usual" scenario–that is without climate action—the industry's global emissions would triple their baseline emissions estimated in 2014 by 2050.

Progressing with the decarbonization of the economic sectors vital to health care will contribute partially towards countries fulfilling their commitments under the Paris Agreement, as depicted in the top wedge. However, even if all Nationally Determined Contributions submitted up to 2017 are achieved (requiring active engagement from the health sector in advocacy endeavors), a substantial portion of emissions will persist, necessitating interventions originating within the health care industry itself. The proposed road map outlines three interconnected decarbonization routes that the industry must navigate to tackle this issue, namely:

- Pathway 1—Reduce the carbon footprint of delivery, facilities, and operations. Health care delivery and operations form the core of the sector's climate impact. Hospitals and health systems worldwide must address their greenhouse gas emissions and adopt measures aimed at completely eliminating carbon emissions from every facet of delivery and its associated functions, all while preserving and enhancing patient care.
- Pathway 2—Decarbonizing supply chain. Over 70% of the sector's climate impact stems from "Scope 3" emissions, much of which originates in the global supply chain. This supply chain encompasses both Pathway 2 and Pathway 3. Pathway 2 involves reducing direct emissions from grid electricity purchased by health care facilities, as well as addressing the production, packaging, and transportation of products utilized in the health care sector. Health systems can utilize procurement decisions to advocate for the decarbonization of their supply chain while leveraging the sector's collective influence by consolidating purchasing power across nations and borders. At the same time, health care manufacturers and suppliers must take prompt action to transition towards zero emissions.
- Pathway 3—Accelerating decarbonization efforts in the broader economy and society. Every component of supply chain and delivery depends on other industries providing energy, chemicals, building materials, packaging, infrastructure, transport, food, and beyond. Comprehensive societal decarbonization is vital for the health care sector to achieve zero emissions, while simultaneously safeguarding the health of both people and the planet against the repercussions of climate change.

As argued by representatives of this initiative, achieving zero emissions necessitates a series of high-impact actions spanning the three pathways. Implementing these actions will lead to a significant reduction in greenhouse gas emissions within the industry.

According to this initiative, these actions ought to consist of: (1) ensuring that industry is powered by zero emissions electricity across the three pathways (blue); (2) ensuring every building and product manufacturing facility and their infrastructure promote energy efficiency, zero emissions, and climate resilience (green); (3) transitioning to 100% low or zero emission fleet vehicles and infrastructure, while encouraging active travel and public transport for patients and staff (pale blue); (4) providing healthy, locally, and sustainably produced fresh and seasonal food with zero food waste (orange); (5) decreasing unnecessary pharmaceutical consumption, replacing high-emission products with environmentally-friendly alternatives, and encouraging the production of affordable green, climate-smart medication (red); (6) incorporating circular economy principles into the procurement of supplies, adoption of clean technologies, reduction of waste volume and toxicity, and sustainable management of waste (lime), and (7) eliminating inefficient and unnecessary practices, integrating carbon reduction with the quality of care, and reinforcing resilience (purple).

## Digital Health

### *What Is It?*

E-health or digital health is a complex term that goes beyond the digital transformation of the health care system. As proposed by Eysenbach (2001): "*It is an emerging field at the intersection of medical informatics, public health and business, referring to health services and information delivered or enhanced through the Internet and related technologies. In a broader sense, the term characterizes not only a technical development, but also a state-of-mind, a way of thinking, an attitude, and a commitment for networked, global thinking, to improve health care locally, regionally, and worldwide by using information and communication technology.*"

Digital health arises from the ongoing digital revolution that is reshaping various industries, including health care, as well as the increasing demand for a more streamlined health care system, reducing bureaucracy, enhancing speed in managing a larger patient base, and improving efficiency.

Digital health comprises a wide range of heterogeneous technologies and services, which have been developed to address different health objectives (i.e., improve the quality of outcomes of care and service, population health, patient experience, provider experience or address health disparities), system challenges (e.g., information, such as delayed reporting of events or cost, such as lack of effective resource allocation), system categories (e.g., client communication systems, emergency response systems), health care providers and clients (WHO, 2018b). These include remote sensing and wearables; telemedicine and health information;

data analytics and intelligence, predictive modeling; health and wellness behavior modification tools; bioinformatics tools; medical social media; digitized health and medical record platforms; patient-physician-patient portals; diagnostics, compliance, and treatments; decision support systems and imaging (Ronquillo et al., 2024).

## Electronic Health Records

Electronic Health Records (EHRs) stand as a cornerstone of digital health. The term EHR was first introduced in 1998 in the UK with the Information for Health white paper (Burns, 1998), which aimed to revitalize the National Health Service information technology, shifting its emphasis from administrative to clinical processes. It established deadlines for linking general practices to the National Health Service network in UK and implementing electronic patient records in secondary care by April 2005. Additionally, the strategy focused on the need to address the "millennium bug" and restructuring the administration's central information technology agencies, in so doing marking the start of an ongoing period of transformation (Cross, 2006).

Several definitions and clarifications regarding electronic records have been proposed in the literature, among which Burns (1998), who stated that "*electronic patient record describes the record of the periodic care provided mainly by one institution. Typically, this will relate to the health care provided to a patient in a hospital. EPRs may also be held by other healthcare providers, for example specialist units or mental health NHS Trust. The term EHR is used to describe the concept of a longitudinal record of a patient's health and healthcare—from cradle to grave. It combines both the information about patient contacts with primary healthcare as well as subsets of information associated with the outcomes of periodic care held in electronic patient records.*"

The ambition of EHRs is to enable citizens to track and access their complete health care history, allowing them to share it with health care professionals for a more effective and efficient service. The nature and content of EHRs may vary across regions and countries, though they typically include the following items: identification and administrative data of the patient, medical reports, first aid minutes, hospital discharge letters, summary health profile, pharmaceutical dossier, and consent or refusal to donate organs and tissues (Watzlaf et al., 2004).

EHRs also comprise: a Patient Summary, that is, a certified document prepared by the general practitioner, and the "personal notebook of the assisted person," which contains personal information not certified by health care operators (Häyrinen et al., 2008). The Patient Summary holds significant importance and serves as the patient's health ID card. It includes comprehensive clinical information describing the patient's status, a list of relevant health-related issues, diagnoses, allergies, pharmacological therapies for potential chronic illnesses, and other necessary details for patient care. The patient's personal notebook is a separate section where patients can independently add information and documentation, they deem valuable for their care, such as data collected from self-monitoring mobile applications.

Focusing on the potential benefits of EHRs (Menachemi & Collum, 2011), they provide unique access to various documents like reports, prescriptions, and resignation letters, hence eliminating the need for patients to carry all paperwork and reducing the risk of document loss or damage. Patients can access their clinical history digitally at any point in time. Furthermore, improved access to all patient history helps avoiding the risk of repeating clinical exams and develops more reliable diagnosis. Indeed, physicians can promptly access patient clinical status information in emergencies, saving time and ensuring correct intervention, thereby safeguarding patient health. With patient consent, medical staff can consult the patient's medical history in digital format, necessary for improving service quality, resulting in shorter waiting times compared to paper-based analysis. Nonetheless, concerns such as the expenses related to adopting and integrating EHR systems, the ongoing maintenance costs, potential revenue loss due to temporary decreases in productivity, disruption of work-flows for medical staff and providers, all present disincentives especially for small and medium-size hospitals and health care providers to adopt and implement EHRs.

## ePrescribing

E-prescribing refers to the electronic generation and transmission of medical prescriptions, eliminating the need for traditional paper prescription sheets (Porterfield et al., 2014). This technology enables doctors to input prescription requests into a management software, including patient details, prescribed medications or tests, and any exemptions. The information is typically stored in a central database accessible to pharmacies, streamlining the prescription process and allowing patients to access their medications from any pharmacy or affiliated health care facility. While some countries, such as like Denmark and Sweden, have fully integrated electronic prescribing systems nationwide, others such as the USA, England, and Finland have adopted it to a lesser extent, each with unique and ad hoc approaches (Aldughayfiq & Sampalli, 2021). Despite the lack of a unified standard, numerous studies have evaluated the potential benefits and barriers of e-prescribing (Gagnon et al., 2014). One significant benefit is the reduction in medication errors resulting from illegible prescriptions or filling mistakes. Studies have shown a substantial decrease in error rates after implementing e-prescribing systems. Moreover, e-prescribing improves efficiency and reduces costs by minimizing paperwork, streamlining communication between prescribers and pharmacies, and saving time on administrative tasks. Estimates suggest substantial cost savings over time due to fewer adverse drug events and reduced visits to healthcare facilities (McMullin et al., 2005).

### Telemedicine

The rapid development of technology has introduced new opportunities in the health care sector to support traditional care methods. In recent years, the use of telemedicine has significantly increased, allowing for easier connection between patients, doctors, and professionals from different locations. Various definitions of telemedicine have been proposed, with one by WHO, which reads: "*Telemedicine is the delivery of health care services, where distance is a critical factor, by all health care professionals using information and communication technologies for the exchange of valid information for diagnosis, treatment and prevention of disease and injuries, research and evaluation, and for the continuing education of health care providers, all in the interests of advancing the health of individuals and their communities.*" (WHO, 1998).

Thus, telemedicine is a method through which health care services can be provided remotely by exchanging medical information. Information is exchanged using communication technologies and may include data, text, audio, images, and video. Telemedicine is not meant to replace traditional methods, but rather complement them by providing new approaches aimed at improving the quality and efficiency of health care. This digital approach is not limited to data exchange or analysis, but also offers opportunities for improving care protocols, as doctors or professionals from different locations can interact and collaborate in real-time to provide suitable care for the patient.

Telemedicine can lend itself to several clinical applications (WHO Global Observatory for eHealth, 2010), including:

- Health emergencies: First responders can directly manage patients who arrive at the emergency room even in the absence of specialists, by contacting doctors and professionals from other hospitals. This allows avoiding time loss in transfers and for intervening promptly. These aspects are particularly important, especially for patients with time-sensitive conditions. To this end, many local authorities are adopting an organizational model called *hub-spoke*, where specialized doctors and professionals are present in hub centers to provide support services such as teleconsultation or specialist telemedicine (see below).
- Accessibility to remote diagnostic services, which enables the provisions of these services even for patients located in remote areas.
- Remote control and monitoring: Telemedicine-supporting devices include functionalities for monitoring patients' vital parameters with the aim of preventing or avoiding the worsening of clinical conditions.
- Chronic conditions: Telemedicine is also useful for managing chronic conditions and relevant clinical cases. It reduces the need for transfers and travel of vulnerable patients, and allows for monitoring patients' vital parameters, which are then analyzed to provide appropriate care.

Telemedicine can cover different types of health care services, including specialist services such as:

- Televisit: This is a medical activity involving a remote visit between the doctor and the patient using video call systems. Televisits may lead to the prescription of medications or treatments.
- Teleconsultation: This is an activity where a professional remotely consults with other doctors to discuss and collaborate on the patient's clinical situation, referring to shared health data and information. Teleconsultation can also be conducted in real-time with the presence of the patient, in which case a multidisciplinary approach is adopted. The goal is to formulate a diagnosis and/or identify a treatment.
- Telecooperation: This is an assistance activity involving medical doctors and other health care professionals (Saleh et al., 2015).
- Telehealth refers to a set of services that enable patient management, especially those with chronic conditions including systems for monitoring the patient's vital parameters. These systems collect and continuously transmit data to be interpreted and analyzed by doctors and specialists.
- Telecare is a set of support services, primarily provided at home, mainly targeting elderly or fragile individuals. It involves the implementation of alarms, emergency systems, devices, and applications for sharing data with health care facilities. Telecare falls more within the realm of social assistance, but it still ensures continuous health care assistance with a preventive purpose.

## M-health and Wearable Devices

M-health has been defined as medical and public health practices supported by mobile devices, such as mobile phones, patient monitoring devices, personal digital assistants, and other wireless devices (Istepanian et al., 2006). Put simply, a mobile health application can be defined as a software application designed to perform health-related tasks and functions. M-health signifies how mobile and wireless technologies can play a vital role in enhancing health-related services. The proliferation of smartphone technology and health services has enabled the development of a multitude of mobile applications to promote people's health. Indeed, the production and use of m-health applications has increased significantly among medical professionals and healthcare consumers in recent times. In 2012, for instance, more than 40,000 health-related applications were already available on smartphone app stores in the USA (West, 2012). And an upward trend is expected to continue (Lu et al., 2020).

The concept of M-health gained traction in 2009 with the formation of the mHealth alliance, aimed at leveraging smartphone technology to improve health care services (Almarri & Bhatti, 2015). Since then, health care applications have proven to be valuable tools for patients in monitoring their health conditions. These applications assist in medication adherence, disease monitoring, and self-accountability. They provide reminders to improve medication adherence and reduce missed appointments. With advancements in artificial intelligence and smartphone technologies, the role of m-health has become increasingly crucial

(Whitehead & Seaton, 2016). These applications have the potential to optimize health care systems worldwide, leading to benefits for both healthcare providers and patients, such as cost reduction, improved access to facilities, and enhanced diagnostic capabilities for physicians. However, discontinuation of e-health and m-health application usage may occur due to factors like lack of realization, need, cost, tech-savviness, and motivation (Peng et al., 2016).

As for devices, the integration of hardware with health care applications offers various benefits, providing health-related data for medical diagnosis and behavior change interventions. Wearable gadgets like smartwatches, sensors, and camera clips can measure vital health parameters and influence positive behaviors, despite some users discontinuing usage due to poor user experience. Wearable gadgets can aid in managing chronic diseases by measuring sleep duration, physical activity, and depressive symptoms. They can positively influence physical activity levels, body mass index, stress levels, and blood pressure (Piwek et al., 2016). A research study has highlighted physicians' expectations regarding health wearable gadgets, which are anticipated to monitor fitness activities, communicate with physicians, and identify risk factors for early treatment and wellness maintenance (Volpato et al., 2021). Perceived advantages encompassed aiding patients in feeling empowered and educated, assisting in behavior change, enhancing awareness of personal medical records and bodily functions, facilitating efficient transmission of information, and fostering connection with the patient's medical network. However, general practitioners have expressed apprehensions regarding the absence of scientific validation, uncertainties regarding data protection, and potential financial interests linked with stakeholders. Additional perceived risks involved the promotion of an excessively medicalized health culture and the possibility of encouraging patients to engage in self-diagnosis and self-medication (Canali et al., 2022).

## *The Rise of Digital Transformation*

The COVID-19 pandemic accelerated the digital transformation of numerous services, particularly in health care, where there has been a rapid increase in the adoption of new services and products. Consequently, a multitude of online platforms have emerged to cater to the evolving needs of patients. However, this process started well before the pandemic across Western countries (Baudier et al., 2023). A cornerstone of the digitalization process across European state members is the Digital Agenda for Europe, in its national and local declinations (European Parliament, 2023). Building upon the Lisbon strategy, the 2010 Digital Agenda for Europe emphasized the significance of Information and Communication Technologies (ICTs) in achieving the EU's goals. In 2015, the digital single market strategy expanded on the Digital Agenda, focusing on three main pillars: (1) enhancing access to digital goods and services throughout Europe; (2) creating favorable conditions for digital networks and services; and (3) maximizing the growth potential of the digital economy (European Commission, 2015).

The 2020 strategy, geared towards shaping Europe's digital future, focused on technologies that would serve the interests of individuals, foster a competitive economy, and uphold an open, democratic society. In 2021, this strategy was further enriched with the introduction of the Digital Compass for 2030 (European Commission, 2021a), outlining the EU's digital objectives for the coming decade.

Following the onset of the COVID-19 pandemic, unveiled on March 9, 2021 the EU Digital compass outlined four key targets to achieve by 2030, namely:

- Skills: Ensuring that at least 80% of adults possess basic digital skills, with 20 million ICT specialists employed in the EU, and a concerted effort to increase female representation in these roles.
- Businesses: Facilitating the adoption of cloud-computing services, big data, and AI by 75% of companies, while ensuring that over 90% of small and medium-sized enterprises in the EU attain a basic level of digital proficiency. Additionally, the aim is to double the number of EU-based unicorns.
- Infrastructure: Ensuring gigabit connectivity for all EU households, 5G coverage for all populated areas, and striving for Europe to contribute 20% of global semiconductor production, focusing on sustainability. The deployment of 10,000 climate-neutral highly secure edge nodes and the establishment of Europe's first quantum computer are also objectives.
- Public services: Making all key public services available online, granting citizens access to their e-medical records, and aiming for 80% of citizens to utilize electronic identity solutions.

The Digital Europe Programme, introduced through Regulation (EU) 2021/694 (European Union, 2021), is an EU initiative allocating €7.5 billion (2021–2027) to digital technology projects, including supercomputing, AI, cybersecurity, advanced digital skills, and digital tech integration. These efforts are supported by digital innovation hubs and will be coordinated with other EU funds such as Horizon Europe (European Commission, 2021b) and the Recovery and Resilience Facility (European Commission, 2022). Member States are required to allocate at least 20% of their recovery funds towards digitalization projects during the COVID-19 recovery, as mandated by Regulation (EU) 2021/694.

## *Opportunities and Challenges*

Action plans and strategies (WHO, 2022; European Health and Digital Executive Agency, 2021) have been designed to support countries in leveraging and scaling up the digital transformation discussed above, which has boosted the development of new health services, products, and devices. Digital health technologies are frequently promoted as means to reduce the environmental impact of health care systems without compromising the quality and financial sustainability of services (Thompson, 2021). However, whether the adoption of these technologies can help mitigate the issue of GHG emissions remains uncertain for most instances. For example, the use of Artificial Intelligence for medical imaging. While this digital technology has revolutionized medical imaging, aiding tasks like breast cancer screening and tuberculosis detection, there are concerns that its computational needs could exacerbate the carbon footprint. For some other digital technologies, instead, evidence suggests positive effects. This is the case of telemedicine. In fact, reducing the number of in-person consultations has been found as an effective means of lowering health-care emissions, by reducing the need for patient and provider travel (The Lancet Digital Health, 2023).

In other instances, there will be trade-offs between pollutants. This is the case of digital technologies for transitioning the current waste management, for the health care industry and beyond, towards a more sustainable approach to handling materials. These technologies enhance recycling processes, making it easier for producers to utilize recyclable materials, enabling consumers to make informed purchasing and sorting choices, and providing recyclers with improved options for sourcing waste materials. However, the benefits reaped from more technologically advanced waste management systems will be compensated by an increased computational demand and therefore GHG emissions.

In addition, the current pace of digital health innovation is likely widening the so-called Digital Divide (Makri, 2019). WHO emphasizes equity as a crucial aspect of eHealth, underscoring the importance of adopting a social justice approach to the development and planning of these initiatives. In this context, equity should be prioritized over equality. However, existing levels of inequality pose challenges in assessing the impact of digital health, understanding its significance, and creating scalable solutions in underserved communities. The advent of the Covid-19 pandemic has prompted a significant shift in the perception and adoption of digital health solutions. The pandemic coincided with a time when technology could enhance or substitute traditional healthcare methods, presenting opportunities for advancements in healthcare through big data analysis and data sharing tools Batko & Ślęzak, 2022). These advancements have provided a broader platform for digital health interventions, such as telehealth, online support systems, and improvements in supply chain management, to reach individuals who have limited access to healthcare services. They have also facilitated progress in medical research and increased efficiencies in the healthcare supply chain. However, a critical question remains on whether these advancements effectively contributed to bridging the health equity gap or, on the contrary, exacerbated the existing disparities in health care access (Hadjiat, 2023).

## ICT

The use of innovative information and communication technologies can reduce care waiting time for medical assistance, which is especially crucial in emergency cases. It can also enable real-time monitoring of vital signs and health conditions of patients, improving the quality of care and therapy. The process of digital transformation removes the geographical limitation of access-related services (Al-Shorbaji,

2021). Digital health can also ensure continuity of care, leading to service improvement and potentially creating stimulating opportunities for businesses operating in the sector. However, these technologies present some critical aspects and challenges related to organizational and infrastructural inadequacies, fueling forms of discrimination among different territorial contexts and/or within the same geographic area due to varying levels of technological literacy among users (Imran, 2023).

From a medical standpoint, one of the opportunities associated with digital health is to expedite the implementation of medical research findings into health care practices, involving all relevant fields from diagnostics and therapies to technological devices and decision-making processes (Stoumpos et al., 2023). These expectations involve the prospect of enhancing the efficiency of both individual health care professionals and public health care administration. It is noteworthy that the use of robotics in surgical fields has long expanded the boundaries of medical-human efficiency, enhancing and sometimes even extending individual knowledge and skills, allowing for maneuvers otherwise unfeasible. The perspectives offered by new digitalization techniques currently concern both the modes of the doctor–patient relationship (which, for example, can increasingly rely on telemedicine) and the potential (or rather, progressive) replacement, in favor of artificial intelligence, in diagnostic medicine (Győrffy et al., 2020). As for telemedicine, international legal reflections have highlighted the issue of effective access to this tool, revealing how, far from being a vector of inevitable social leveling, it can exacerbate substantive inequalities, both geographically and demographically (e.g., if it becomes de facto predominantly the prerogative of the young and more digitized users, thus disadvantaging the elderly population, which most needs efficient and constant access to healthcare; Haimi, 2023).

Regarding the promise of digital health care improvement in supporting the system, reference must be made to the proposed use of digital technologies as a tool for reducing (or eliminating) corruption, for example, through the application of blockchain technologies to public contracts. In particular, international studies have highlighted how the use of digital technologies that improve the visibility and understanding of administrative action can be an effective tool in the fight against corruption by increasing transparency and accountability on the part of users (Mackey & Cuomo, 2020; Kaboré et al., 2022). In addition to these issues, the international debate has focused on the problem of personal data protection. It is immediately worth noting the close correlation between data access and the effectiveness of digital healthcare systems, especially those of predictive diagnostic medicine.

An issue has been highlighted, called universal *datification*, that is, the collection—and use—of data, even if not strictly health-related (Trocin et al., 2023). This has profound implications for privacy, as it requires a redefinition of what constitutes private data. The proper balance between legal protection of data and perspectives of digital health care efficiency transcends the scope of medical practice and the efficiency of health care administration. Closely related is the so-called universal *valorization* of data (Ho, 2022), especially by and through private actors, a circumstance that further blurs the already rather tenuous line between public power and private (economic) initiative. It is indeed interesting to note that, from the user's point of view, transparency and access to one's own data have become the default option, while the related protection is presented as an additional service. From this perspective, the growing commercialization of health data risks reinforcing social and economic divisions, threatening the constitutional obligation to protect substantive equality, and the very effectiveness of digital healthcare.

The paradigm shift imposed by digitalization strategies in the face of organizational and structural issues requires a reflection on the allocation of responsibilities. In an era characterized by the rapid evolution of technology and the progressive expansion of the areas in which it is used, the uniqueness of the health care scenario requires not to underestimate the concepts of dignity and identity of the patient (Brall et al., 2019). The ongoing digital transition, besides impacting the ways in which care and assistance activities are carried out and requiring a revision of the organizational framework supporting them, prompts a reflection on the mechanisms of attribution of the choices made, as well as the borne responsibility by structures and professionals in the provision of services, especially when performed through artificial intelligence techniques (Bleher & Braun, 2022). Indeed, the integration of artificial intelligence in health care places health protection issues in the context of modern risk, consequently complicating the process of identifying potential technological hazards that may impact the right to health, as well as establishing rules and parameters on which to build a health care relationship (Karni et al., 2020). Technology has sometimes been considered the cause of the distance between doctor and patient and, therefore, of the depersonalization of health care relationships where, instead, direct contact is traditionally considered an element of connection between the legal and ethical duty incumbent on the former to provide care (and assistance) and the effectiveness of the latter's right to receive it. Moreover, digital technology has been seen as an obstacle to tracing the professional's work with negative consequences both on the effectiveness of treatments and treatments, and on demonstrating the respective legal responsibility of those who use it, making its boundaries uncertain (Borges do Nascimento et al., 2023).

The advancement of new technologies has implied an enrichment of the profiles and sector-specific skills required to professionals, resulting in a process of specialization in medicine. As a result, not only is medical and health care activities required to conform to increasingly high-quality standards, but the provision of such services is often the result of the contribution of a plurality of individuals with specific skills, with implications for the possible reshaping of the concept of professional diligence. The qualified professional position imposes and justifies the assumption of a warranty obligation towards the recipient of the service, generating in the latter an expectation of the correct and timely application of the specific technical rules that oversee the performance of the activity itself. According to this paradigm, therefore, the doctor's responsibility stems from a 'sui generis' obligation defined as without a primary obligation of performance, which materializes whenever the doctor has prejudiced the trust placed by the patient in their professionalism. The debate on the contours of healthcare responsibility has certainly been enlivened by the overcoming of the distinction between obligations of means and obligations of results (Silven et al., 2022). That said, it remains to be determined whether such theoretical elaborations, subject to readjustment in the context of healthcare responsibility, are destined for further adjustments when they concern services and performances rendered in the same context, but with technologically innovative methods.

## References


Alderwick, H., Hutchings, A., Briggs, A., et al. (2021). The impacts of collaboration between local health care and non-health care organizations and factors shaping how they work: A systematic review of reviews. *BMC Public Health, 21*, 753. https://doi.org/10.1186/s12889-021-10630-1

Aldughayfiq, B., & Sampalli, S. (2021). Digital health in physicians' and pharmacists' office: A comparative study of e-prescription systems' architecture and digital security in eight countries. *OMICS: A Journal of Integrative Biology, 25*(2), 102–122.

Almarri, A., & Bhatti, T. (2015). Consumers' attitude towards the use of mobile health apps: An empirical review. In *The second international conference on Electrical and Electronics Engineering, Clean Energy and Green Computing (EEECEGC2015)* (pp. 56–63). Society of Digital Information and Wireless Communication at Mevlana Universitesi.

Al-Shorbaji, N. (2021). Improving healthcare access through digital health: The use of information and communication technologies. In A. Agrawal & S. Kosgi (Eds.), *Healthcare access* (pp. xx–xx). IntechOpen. https://doi.org/10.5772/intechopen.99607

Amoo, N. A., Malby, R., & Mervyn, K. (2016). Innovation and sustainability in large-scale healthcare improvement collaborative–seven propositions for achieving system-wide innovation and sustainability. *International Journal of Sustainable Strategic Management, 5*(2), 149–179. https://doi.org/10.1504/IJSSM.2016.080465

Arup. (2024). *Climate and sustainability services*. Accessed March 05, 2024, from https://www.arup.com/services/climate-and-sustainability-services

Basel Convention. (2020). *Basel convention technical guidelines*. Accessed March 06, 2024, from https://www.basel.int/?tabid=8025

Batko, K., & Ślęzak, A. (2022). The use of Big Data Analytics in healthcare. *Journal of Big Data, 9*(1), 3. https://doi.org/10.1186/s40537-021-00553-4

Baudier, P., Kondrateva, G., Ammi, C., Chang, V., & Schiavone, F. (2023). Digital transformation of healthcare during the COVID-19 pandemic: Patients' teleconsultation acceptance and trusting beliefs. *Technovation, 120*, 102547. https://doi.org/10.1016/j.technovation.2022

Blackstone, E. A., & Fuhr, J. P., Jr. (2007). Biopharmaceuticals: The economic equation. *Biotechnology Healthcare, 4*(6), 41–45.

Bleher, H., & Braun, M. (2022). Diffused responsibility: Attributions of responsibility in the use of AI-driven clinical decision support systems. *AI and Ethics, 2*(4), 747–761. https://doi.org/10.1007/s43681-022-00135-x

Booth, A., Jager, A., Faulkner, S. D., Winchester, C. C., & Shaw, S. E. (2023). Pharmaceutical company targets and strategies to address climate change: Content analysis of public reports from 20 pharmaceutical companies. *International Journal of Environmental Research and Public Health, 20*(4), 3206. https://doi.org/10.3390/ijerph20043206

Borges do Nascimento, I. J., Abdulazeem, H. M., Vasanthan, L. T., Martinez, E. Z., Zucoloto, M. L., Østengaard, L., Azzopardi-Muscat, N., Zapata, T., & Novillo-Ortiz, D. (2023). The global effect of digital health technologies on health workers' competencies and health workplace: An umbrella review of systematic reviews and lexical-based and sentence-based meta-analysis. *The Lancet Digital Health, 5*(8), e534–e544. https://doi.org/10.1016/S2589-7500(23)00092-4

Brall, C., Schröder-Bäck, P., & Maeckelberghe, E. (2019). Ethical aspects of digital health from a justice point of view. *European Journal of Public Health, 29*(Supplement_3), 18–22. https://doi.org/10.1093/eurpub/ckz167

Burns, F. (1998). *Information for health: An information strategy for the modern NHS. 1998–2005*. Department of Health.
Canali, S., Schiaffonati, V., & Aliverti, A. (2022). Challenges and recommendations for wearable devices in digital health: Data quality, interoperability, health equity, fairness. *PLOS Digital Health, 1*(10), e0000104. https://doi.org/10.1371/journal.pdig.0000104
Cross, M. (2006). Will connecting for health deliver its promises? *BMJ, 332*(7541), 599–601. https://doi.org/10.1136/bmj.332.7541.599
European Commission. (2015). *A digital single market strategy for Europe*. Accessed March 06, 2024, from https://eur-lex.europa.eu/legal-content/EN/TXT/?uri=celex%3A52015DC0192
European Commission. (2021a). *2030 Digital Compass: The European way for the Digital Decade*. Accessed March 06, 2024, from https://eur-lex.europa.eu/legal-content/en/TXT/?uri=CELEX%3A52021DC0118
European Commission. (2021b). *Research and innovation: Horizon Europe*. Accessed March 06, 2024, from https://research-and-innovation.ec.europa.eu/funding/funding-opportunities/funding-programmes-and-open-calls/horizon-europe_en
European Commission. (2022). *REPowerEU Plan*. Accessed March 06, 2024, from https://eur-lex.europa.eu/legal-content/EN/TXT/?uri=COM%3A2022%3A230%3AFIN&qid=1653033742483
European Health and Digital Executive Agency (HaDEA). (2021). *About HaDEA*. Accessed March 06, 2024, from https://hadea.ec.europa.eu/about_en
European Parliament. (2023). *Digital agenda for Europe*. Accessed March 06, 2024, from https://www.europarl.europa.eu/factsheets/en/sheet/64/digital-agenda-for-europe
European Union. (2021). *Regulation (EU) 2021/694 of the European Parliament and of Council of 29 April 2021 establishing the Digital Europe Programme and repealing Decision (EU) 2015/2240*. Accessed March 06, 2024, from https://eur-lex.europa.eu/legal-content/EN/TXT/?uri=CELEX%3A32021R0694
Eysenbach, G. (2001). What is e-health? *Journal of Medical Internet Research, 3*(2), e20. https://doi.org/10.2196/jmir.3.2.e20
Gagnon, M. P., Nsangou, É. R., Payne-Gagnon, J., Grenier, S., & Sicotte, C. (2014, May–June). Barriers and facilitators to implementing electronic prescription: A systematic review of user groups' perceptions. *Journal of the American Medical Informatics Association, 21*(3), 535–541. https://doi.org/10.1136/amiajnl-2013-002203
GHG Protocol. (2024). *About us*. Accessed March 06, 2024, from https://ghgprotocol.org/about-us
Győrffy, Z., Radó, N., & Mesko, B. (2020). Digitally engaged physicians about the digital health transition. *PLoS One, 15*(9), e0238658. https://doi.org/10.1371/journal.pone.0238658
Hadjiat, Y. (2023). Healthcare inequity and digital health–A bridge for the divide, or further erosion of the chasm? *PLOS Digital Health, 2*(6), e0000268. https://doi.org/10.1371/journal.pdig.0000268
Haimi, M. (2023). The tragic paradoxical effect of telemedicine on healthcare disparities – A time for redemption: A narrative review. *BMC Medical Informatics and Decision Making, 23*(1), 95. https://doi.org/10.1186/s12911-023-02194-4
Häyrinen, K., Saranto, K., & Nykänen, P. (2008). Definition, structure, content, use and impacts of electronic health records: A review of the research literature. *International Journal of Medical Informatics, 77*(5), 291–304. https://doi.org/10.1016/j.ijmedinf.2007.09.001
Health Care Without Harm. (2019). *Climate-smart health care series, Green Paper Number One*. Accessed March 06, 2024, from https://noharm-global.org/sites/default/files/documents-files/5961/HealthCaresClimateFootprint_092319.pdf
Health Care Without Harm. (2021a). *Global road map for health care decarbonization*. Accessed March 06, 2024, from https://healthcareclimateaction.org/roadmap
Health Care Without Harm. (2021b). *Climate-smart health care series, Green Paper Number Two*. Accessed March 06, 2024, from chrome-extension://efaidnbmnnnibpcajpcglclefindmkaj/https://healthcareclimateaction.org/sites/default/files/2021-06/Health%20Care%20Without%20Harm_Health%20Care%20Decarbonization_Road%20Map.pdf

Health Protection Agency. (2010). *The impact on health of emissions to air from municipal waste incinerators*. [Online]. Retrieved from http://www.plymouth.gov.uk/hpa_impact_on_health_of_emissions_to_air.pdf

Ho, C. W. (2022). Operationalizing "one health" as "one digital health" through a global framework that emphasizes fair and equitable sharing of benefits from the use of artificial intelligence and related digital technologies. *Frontiers in Public Health, 10*, 768977. https://doi.org/10.3389/fpubh.2022.768977

IHME (Institute for Health Metrics and Evaluation). (2021). *COVID-19 projections*. Accessed March 06, 2024, from https://repository.gheli.harvard.edu/repository/13194/

Imran, A. (2023). Why addressing digital inequality should be a priority. *The Electronic Journal of Information Systems in Developing Countries, 89*(3), e12255. https://doi.org/10.1002/isd2.12255

Istepanian, R. S. H., Laxminarayan, S., & Pattichis, C. S. (Eds.). (2006). *M-Health: Emerging mobile health systems. Topics in biomedical engineering*. Springer. https://doi.org/10.1007/b137697

Kaboré, S. S., Ngangue, P., Soubeiga, D., Barro, A., Pilabré, A. H., Bationo, N., Pafadnam, Y., Drabo, K. M., Hien, H., & Savadogo, G. B. L. (2022). Barriers and facilitators for the sustainability of digital health interventions in low and middle-income countries: A systematic review. *Frontiers in Digital Health, 4*, 1014375. https://doi.org/10.3389/fdgth.2022

Karni, L., Dalal, K., Memedi, M., Kalra, D., & Klein, G. O. (2020). Information and communications technology-based interventions targeting patient empowerment: Framework development. *Journal of Medical Internet Research, 22*(8), e17459. https://doi.org/10.2196/17459

Lenzen, M., Malik, A., Li, M., Fry, J., Weisz, H., Pichler, P.-P., Su-veges Moreira Chaves, L., Capon, A., & Pencheon, D. (2020). The environmental footprint of health care: A global assessment. *The Lancet Planetary Health, 4*(7), e271–e279. https://doi.org/10.1016/S2542-5196(20)30121-2

Liu, M., Li, H., Song, A., Peng, P., Liu, H., Hu, J., Sheng, G., & Ying, G. (2023). Polybrominated dibenzo-p-dioxins/furans and their chlorinated analogues in sediments from a historical hotspot for both brominated flame retardants and organochlorine pesticides. *Environmental Pollution, 316*(Part 1), 120489. https://doi.org/10.1016/j.envpol.2022.120489

Lu, L., Zhang, J., Xie, Y., Gao, F., Xu, S., Wu, X., & Ye, Z. (2020). Wearable health devices in health care: Narrative systematic review. *JMIR mHealth and uHealth, 8*(11), e18907. https://doi.org/10.2196/18907

Macassa, G., & Tomaselli, G. (2020). Rethinking developed nations' health systems through a social sustainability perspective in the light of the COVID-19 pandemic: A viewpoint. *Journal of Public Health Research, 9*(4), 1834. https://doi.org/10.4081/jphr.2020

Mackey, T. K., & Cuomo, R. E. (2020). An interdisciplinary review of digital technologies to facilitate anti-corruption, transparency and accountability in medicines procurement. Global Health Action, 13(sup1), https://doi.org/10.1080/16549716.2019.1695241

Maghsoudi, T., Cascón-Pereira, R., & Hernández Lara, A. B. (2020). The role of collaborative healthcare in improving social sustainability: A conceptual framework. *Sustainability, 12*(8), 3195. https://doi.org/10.3390/su12083195

Makri, A. (2019). Bridging the digital divide in health care. *The Lancet Digital Health, 1*(5), e204–e205. https://doi.org/10.1016/S2589-7500(19)30111-6

Mariani, L., Trivellato, B., Martini, M., et al. (2022). Achieving sustainable development goals through collaborative innovation: Evidence from four European initiatives. *Journal of Business Ethics, 180*, 1075–1095. https://doi.org/10.1007/s10551-022-05193-z

McMullin, S. T., Lonergan, T. P., & Rynearson, C. S. (2005). Twelve-month drug cost savings related to use of an electronic prescribing system with integrated decision support in primary care. *Journal of Managed Care Pharmacy, 11*(4), 322–332. https://doi.org/10.18553/jmcp.2005.11.4.322

Menachemi, N., & Collum, T. H. (2011). Benefits and drawbacks of electronic health record systems. *Risk Management and Healthcare Policy, 4*, 47–55. https://doi.org/10.2147/RMHP.S12985

One Planet Summit. (2023). *Coming together to act for the planet*. Accessed March 02, 2024, from https://oneplanetsummit.fr/en

Peng, W., Kanthawala, S., Yuan, S., et al. (2016). A qualitative study of user perceptions of mobile health apps. *BMC Public Health, 16*, 1158. https://doi.org/10.1186/s12889-016-3808-0

Penga, Y., Wua, P., Schartup, A., & Zhang, Y. (2021). Plastic waste release caused by COVID-19 and its fate in the global ocean. *Proceedings of the National Academy of Sciences of the United States of America, 118*, e2111530118.

Pichler, P., Jaccard, I., Weisz, U., & Weisz, H. (2019). International comparison of health care carbon footprints. *Environmental Research Letters, 14*(6), 064004.

Piwek, L., Ellis, D. A., Andrews, S., & Joinson, A. (2016, February 2). The rise of consumer health wearables: Promises and barriers. *PLoS Medicine, 13*(2), e1001953. https://doi.org/10.1371/journal.pmed.1001953

Porterfield, A., Engelbert, K., & Coustasse, A. (2014). Electronic prescribing: Improving the efficiency and accuracy of prescribing in the ambulatory care setting. *Perspectives in Health Information Management, 11*(Spring), 1g. PMID: 24808808; PMCID: PMC3995494.

Ronquillo, Y., Meyers, A., & Korvek, S. J. (2024). Digital Health. In *StatPearls*. Accessed March 06, 2024, from https://www.ncbi.nlm.nih.gov/books/NBK470260/

Rushton, L. (2003). Health hazards and waste management: Impact of environmental pollution on health: Balancing risk. *Environmental Health Perspectives, 68*(1), 183–197.

Saleh, K., Ammon, D., Lehnert, S., Röhr, S., Detschew, V., Specht, M., Henkel, A., & Kaeding, A. (2015). An IHE-conform telecooperation platform supporting the treatment of dementia patients. *Current Directions in Biomedical Engineering*. https://doi.org/10.1515/cdbme-2015-0077

Silven, A. V., van Peet, P. G., Boers, S. N., Tabak, M., de Groot, A., Hendriks, D., van Os, H. J. A., Bonten, T. N., Atsma, D. E., de Graaf, T. J., Sombroek, M. P., Chavannes, N. H., & Villalobos-Quesada, M. (2022). Clarifying responsibility: Professional digital health in the doctor-patient relationship, recommendations for physicians based on a multi-stakeholder dialogue in the Netherlands. *BMC Health Services Research, 22*(1), 129. https://doi.org/10.1186/s12913-021-07316-0

Stoumpos, A. I., Kitsios, F., & Talias, M. A. (2023). Digital transformation in healthcare: Technology acceptance and its applications. *International Journal of Environmental Research and Public Health, 20*(4), 3407. https://doi.org/10.3390/ijerph20043407

Sustainable Markets Initiative. (2024a). *Building a Sustainable Future*. Accessed March 01, 2024, from https://www.sustainable-markets.org/

Sustainable Markets Initiative. (2024b). *Terra Carta*. Accessed March 01, 2024, from https://www.sustainable-markets.org/terra-carta/

Tennison, I., Roschnik, S., Ashby, B., Boyd, R., Hamilton, I., Oreszczyn, T., Owen, A., Romanello, M., Ruyssevelt, P., Sherman, J. D., Smith, A. Z. P., Steele, K., Watts, N., & Eckelman, M. J. (2021). Health care's response to climate change: A carbon footprint assessment of the NHS in England. *The Lancet Planetary Health, 5*(2), e84–e92. https://doi.org/10.1016/S2542-5196(20)30271-0

The Lancet Digital Health. (2023). Curbing the carbon footprint of health care. *The Lancet Digital Health, 5*(12), e848. https://doi.org/10.1016/S2589-7500(23)00229-7

Thompson, M. (2021). The environmental impacts of digital health. *Digital Health, 7*. https://doi.org/10.1177/20552076211033421

Trocin, C., Mikalef, P., Papamitsiou, Z., et al. (2023). Responsible AI for digital health: A synthesis and a research agenda. *Information Systems Frontiers, 25*, 2139–2157. https://doi.org/10.1007/s10796-021-10146-4

United Nations Development Programme. (2023). *Benchmarking health care waste management and DRR capacities in five Asian countries*. Accessed March 06, 2024, from https://www.undp.org/asia-pacific/publications/benchmarking-health-care-waste-management-and-drr-capacities-five-asian-countries

Volpato, L., Del Río Carral, M., Senn, N., & Santiago Delefosse, M. (2021). General practitioners' perceptions of the use of wearable electronic health monitoring devices: Qualitative analysis of risks and benefits. *JMIR mHealth and uHealth, 9*(8), e23896. https://doi.org/10.2196/23896

Watzlaf, V. J., Zeng, X., Jarymowycz, C., & Firouzan, P. A. (2004). Standards for the content of the electronic health record. *Perspectives in Health Information Management, 1*, 1.

West, D. M. (2012). How mobile devices are transforming healthcare issues in technology innovation. *Issues in Technology Innovation, 18*, 1–14. Accessed March 18, 2024, from https://www.brookings.edu/articles/how-mobile-devices-are-transforming-healthcare/

Whitehead, L., & Seaton, P. (2016). The effectiveness of self-management mobile phone and tablet apps in long-term condition management: A systematic review. *Journal of Medical Internet Research, 18*(5), e97. https://doi.org/10.2196/jmir.4883

WHO (World Health Organization). (1998). *A health telematics policy in support of WHO's Health-For-All strategy for global health development: report of the WHO group consultation on health telematics, 11–16 December, Geneva, 1997*. Geneva, Switzerland: World Health Organization.

WHO (World Health Organization). (2015). *Safe health-care waste management: Policy paper*. Accessed March 04, 2024, from http://www.who.int/water_sanitation_health/medicalwaste/hcwmpolicy/en/

WHO (World Health Organization). (2018a). *Health-care waste*. Accessed March 06, 2024, from https://www.who.int/news-room/fact-sheets/detail/health-care-waste

WHO (World Health Organization). (2018b). *Classification of Digital Health Interventions v1.0*. Accessed March 06, 2024, from https://iris.who.int/bitstream/handle/10665/260480/WHO-RHR-18.06-eng.pdf?sequence=1

WHO (World Health Organization). (2022). *Regional digital health action plan for the WHO European Region 2023–2030*. Accessed March 06, 2024, from https://iris.who.int/bitstream/handle/10665/360950/72wd05e-DigitalHealth-220529.pdf?sequence=2 https://www.who.int/docs/default-source/documents/gs4dhdaa2a9f352b0445bafbc79ca799dce4d.pdf

WHO (World Health Organization). (2024). *Alliance for Transformative Action on Climate and Health (ATACH)*. Accessed March 06, 2024, from https://www.who.int/initiatives/alliance-for-transformative-action-on-climate-and-health#:~:text=The%20Alliance%20for%20Transformative%20Action,agenda%20forward%20at%20pace%20and

WHO Global Observatory for eHealth. (2010). *Telemedicine: Opportunities and developments in member states: Report on the second global survey on eHealth*. Accessed March 18, 2024, from https://iris.who.int/bitstream/handle/10665/44497/9789241564144_eng.pdf?sequence=1

Zikhathile, T., Atagana, H., Bwapwa, J., & Sawtell, D. (2022). A review of the impact that healthcare risk waste treatment technologies have on the environment. *International Journal of Environmental Research and Public Health, 19*(19), 11967. https://doi.org/10.3390/ijerph191911967

# Chapter 3
# Normative and Market Instruments of Sustainability

## The Journey of Sustainability Disclosure

The practice of corporate environmental and social reporting has deep roots in the past (Carroll, 2008). Social controversies and environmental concerns have been at the forefront of the public debate for centuries, yet the corporate's formal process for reporting on these issues only started in the early days of modern society (Buhr, 2007). According to Rimmel (2020), the accounting of sustainability and its standardization has undergone a gradual process spanning just over a century, starting with employee reporting, then progressing to social reporting, environmental reporting, and finally sustainability reporting.

Initial forms of social reporting, such as employee reports were already in existence between the late nineteenth and early twentieth centuries. They focused on employee conditions, covering topics such as worker housing, workplace safety, and community development (Mathews, 1995). This practice included the reporting of the social performance of business activities, where in some countries such as France companies with more than 750 employees were required to produce a report on employees—the so-called *bilan social*. Many large companies in other countries would voluntarily produce employee reports, and some also created more comprehensive social reports focusing on corporate citizenship as well as employment (Bennett & James, 1999). However, according to Epstein (1996), these initiatives failed to test of time because had been too extensively used more as a public relations tool than as a means to make fundamental changes in corporate culture.

An ever growing sentiment of distrust in the ability of corporations to meet public expectations and an overall low confidence in the business sector due to a series of environmental disasters (oil spill off the coast of Santa Barbara in 1969, Chernobyl, and Exxon Valdez) occurred throughout the 60s, 70s, and 80s, reached a critical juncture with the signature of international agreements such as the



P. Candio, *Sustainability and Corporate Performance in Health Care*, CSR, Sustainability, Ethics & Governance,
https://doi.org/10.1007/978-3-031-63985-2_3

Montreal Protocol and the Brundtland Report, which stimulated greater attention to the environmental impact of human activities (Laine et al., 2021). The inception of sustainability reporting hence saw the introduction of environmental accounting as its primary methodology (Elkington, 1993). Following the Exxon Valdez incident in 1989, there was a heightened emphasis on environmental reporting, as a means to informing stakeholders about the environmental effects of a company's operations, particularly gaining significance among investors.

Out of the United Nations Conference on Environment and Development in 1992 (Rio de Janeiro), a consensus emerged to establish the Sustainable Development Commission which has since been dedicated to advancing sustainability measurement efforts. This led to the introduction of the so-called Triple Bottom Line approach (Elkington, 1998, 1999). This framework emphasizes the necessity of evaluating economic and business development from three perspectives: people, planet, and profit. Therefore, society expects and demands organizations to report on their role in sustainable development, within a framework of accountability. Building on these steps, in Europe, the EU Directive 2014/95/EU (European Union, 2014)—the so-called the non-financial reporting directive—mandated large companies of public interest in Europe to disclose non-financial and diversity information in their reports starting from 2017, hence underscoring the European Parliament's recognition of the importance of disclosing non-financial information in promoting CSR.

In November 2022, as the regulation currently in force at the time of writing, the European Parliament and the Council of the European Union endorsed the Corporate Sustainability Reporting Directive (CSRD, European Union, 2022), mandating European companies to disclose a range of information concerning the sustainability-related risks and impacts of their business activities. That is when the formal shift from non-financial to sustainability reporting occurred. The CSRD directive introduces more detailed transparency obligations regarding companies' environmental impact, human rights, and social standards. Additionally, it extends the scope of entities subject to sustainability reporting requirements. Notably, not only large public-interest entities (with over 500 employees) already subject to the directive on non-financial reporting will be obliged, but also large enterprises currently not under this directive (with over 250 employees and/or €40 million turnover and/or €20 million total assets), listed small and medium-sized enterprises (excluding microenterprises), small and non-complex credit institutions, captive insurance and reinsurance undertakings, and all third-country entities with EU sales and revenues exceeding €150 million and having at least one subsidiary or branch in the EU. This regulation is set to be implemented in four phases:

1. For financial years commencing on January 1, 2024, disclosure for the 2024 financial year will be mandatory for large public-interest entities (with over 500 employees) already subject to the directive on non-financial reporting.
2. For financial years commencing on January 1, 2025, disclosure for the 2025 financial year will be mandatory for large enterprises currently not subject to the directive on non-financial reporting (with over 250 employees and/or €40 million turnover and/or €20 million total assets).

3. For financial years commencing on January 1, 2026, disclosure for the 2026 financial year will be mandatory for listed small and medium-sized enterprises (excluding microenterprises), small and non-complex credit institutions, and captive insurance and reinsurance undertakings. SMEs can choose not to participate until 2028, providing adequate justification in the management report.
4. For financial years commencing on January 1, 2028, disclosure for the 2028 financial year will be mandatory for third-country entities with EU sales and revenues exceeding €150 million, if they have at least one subsidiary or branch in the EU surpassing certain thresholds.

This phased approach is expected to make sustainability information collection and sharing the norm for almost 50,000 companies in the EU, compared to approximately 11,700 covered by previous regulations. This directive places significant emphasis on the role of statutory auditors or audit firms, by amending the title of Chap. 8 of the 2013 directive from "audit of accounts" to "audit and assurance of compliance with sustainability reporting," and subsequently through amendments to Directive 2004/109/EC, Directive 2006/43/EC, and Regulation (EU) No. 537/2014. Auditors or audit firms are thus required to provide assurance regarding compliance with the principles set out in the directive, including for third-country enterprises. Their reports must be published alongside the annual financial report.

Furthermore, article 2 of Directive 2022/2464, which amends Directive 2004/109/EC on transparency, mandates the Commission to adopt equivalence mechanisms for accounting and sustainability reporting principles to enhance harmonization and comparability of information. Consequently, those responsible for the annual financial report must include statements on compliance with sustainability principles within the management report. The new directive, under Article 3, further elaborates provisions regarding statutory auditors by amending Directive 2006/43/EC concerning "statutory audits of annual and consolidated accounts," focusing on the standards applicable to entities conducting sustainability reporting audits. In addition, provisions regarding professional qualification and training have been amended to ensure an adequate level of competency in sustainability reporting matters and to ensure appropriate consistency between financial and sustainability information. By first October 2028, the Commission will adopt delegated acts aimed at "establishing assurance principles aimed at acquiring a reasonable level of assurance on the compliance of sustainability reporting."

At the time of writing, the latest relevant development on this front is the publication of sustainability standards in the Official Journal of the European Union, following the adoption of the European Sustainability Reporting Standards (ESRS) by the European Commission on 31st July 2023 (European Union, 2023). These standards have been developed in collaboration with The European Financial Reporting Advisory Group (EFRAG) and are divided into three categories: Universal (ESRS 1 and ESRS 2); Thematic (environmental, social, and governance), and Sector-specific standards. All companies, regardless of their sector, must adopt the universal and thematic principles. However, if a company identifies an "impact, risk, or opportunity" not covered by the standards developed by

EFRAG, it is required to comprehensively disclose these impacts, risks, and opportunities. EFRAG has also been tasked with priority development of sector-specific standards relating to the information that must be disclosed by companies operating in the sectors of mining, extractive industries, and coal; oil and gas; road transportation; agriculture, livestock farming, and fishing. The ESRS are designed to provide investors with the necessary insights to gauge the sustainability implications of their invested companies and have been developed while taking into account discussion taken place with the International Sustainability Standards Board and the Global Reporting Initiative to foster extensive interoperability between EU and global standards while mitigating the occurrence of redundant reporting burdens on companies (FISMA, 2023). However, as discussed in the following sections, several reporting frameworks and formats currently exist that are available to corporations and businesses for reporting of impacts and activities relating to sustainability.

## Reporting Formats and Frameworks

When considering the reporting of sustainability, companies are presented with several options, each tailored to specific objectives and audiences. Selecting between an annual report, integrated report, or sustainability report hinges on the company's goals, stakeholders, and the desired depth of information they aim to convey.

### *Annual Report*

All publicly traded companies, and other economic entities such as non-profit organizations and non-governmental institutions, are mandate to prepare and publish an annual report. This document includes five official documents: the balance sheet, the income statement, the cash flow statement, the statement of changes in equity, and the other comprehensive income statement. An annual report represents a summary of the financial position and performance relating to an organization during a financial year and it is one of the main sources of information available to stakeholders (European Commission, 2001a, 2001b). Indeed, this document is instrumental in establishing a relationship of trust and transparency with all interested parties, including investors, and enhancing corporate credibility. At the same time, a report prepared according to criteria of completeness and clarity can also increase attractiveness to potential investors. The annual report is not limited to the mere communication of financial information data, but also plays a role in providing a detailed account of what the company has achieved, and the methods adopted to achieve certain results. This enables analyses of trends and variations in performance indicators, as well as examining relevant market sectors, hence playing a crucial role in informing investment decisions. Nonetheless, preparing an annual report is a costly endeavor in terms of time and financial resources and these reports are often long and complex, which makes understanding by non-specialist users cumbersome (Gilchrist et al., 2023). Moreover, companies often have to comply with regulations regarding the drafting and subsequent disclosure of additional information to be embedded, which could present an additional administrative burden.

Over the last decades, a debate has arisen over the readability of annual reports. Some scholars assert that the intelligibility of these reports can positively impact information asymmetry; however, longer and more complex reports may lead to differing interpretations. On the other hand, other researchers argue that intricate language does not necessarily imply a lack of completeness in the information, as less readable disclosures could be a result of detailed regulations, still providing essential information and helping to mitigate information asymmetry (Panta & Panta, 2023).

Annual reports can deliberately be made difficult to interpret to conceal negative information and hide unfavorable financial scenarios (Qian & Zhu, 2019). A study conducted by Nguyen and Kimura (2023) regarding the association between length of annual reports and information asymmetry highlighted that longer and more detailed annual reports reduce information asymmetry as they provide more information to the public. Over time regulatory authorities have shown an increasing interest in analyzing whether individual investors benefit from clearer and more concise financial information. As regards sustainability, at the time of writing, the annual report can be also used as a means to communicate non-financial information, including on sustainability-related plans and activities. However, empirical studies have found that alternative reporting formats, that is, integrated reporting and sustainability reporting are related to greater accuracy in forecasts compared to an annual report (Rossi & Candio, 2023). In turn, greater accuracy in these reports can lead to more effective capital allocation by investors. Over time, however, recent evidence has showed that listed companies have gradually moved away from this reporting choice in voluntary settings (Fig. 3.1).

## *Materiality Principles*

Before delving into the two formats of reporting alternative to the annual report, let us focus on a key principle underlying the disclosure of sustainability: the materiality principle. This is a broad and pervasive concept, inherently flexible, modified, and realigned over time to address new challenges and informational needs (Edgley et al., 2015). Traditionally, materiality has been defined and observed through the lens of financial reporting. However, today there is consensus that the traditional financial report is not sufficient to provide the necessary information to investors and stakeholders (Saraite-Sariene et al., 2020). Nonetheless, it is difficult for an external analyst to judge whether a particular issue is material for a particular

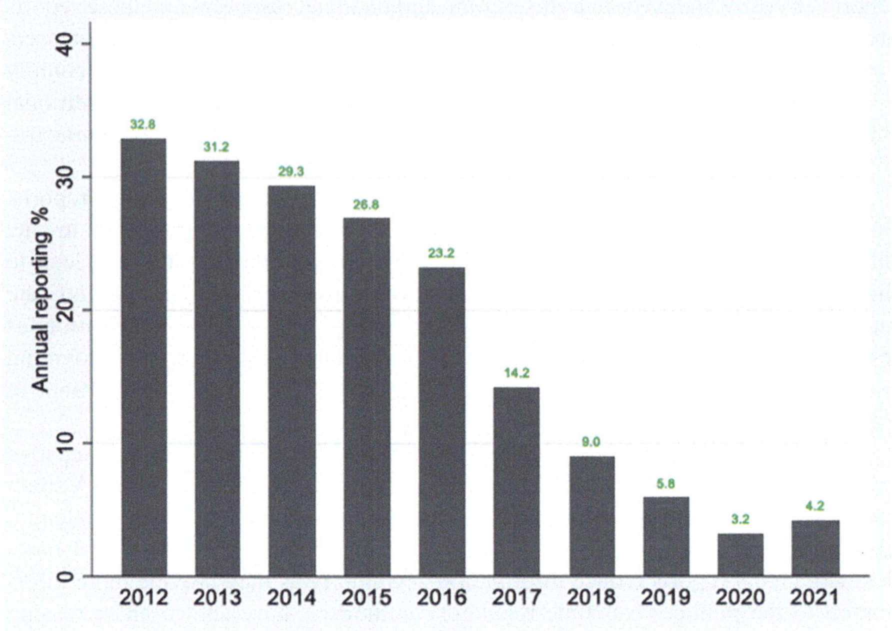


**Fig. 3.1** Trend in the choice of annual reporting. Source: https://pubmed.ncbi.nlm.nih.gov/37659364/

company, and the analysis often depends on the corporate strategy and business model (Mio et al., 2020).

The Directive 2022/2464 focused on the development of standardized reporting models, suitable for overcoming incomparability of reporting format prepared according to different reporting principles. In this regard, the EFRAG, which proceeded with the issuance of Sustainability Reporting Standards, was delegated to that end. The Directive 2022/2464 introduced the principle of double materiality, which—put simply—allows identifying the sustainability aspects most relevant to the company and its stakeholders both in terms of caused and suffered impacts. This concept plays a primary role in the context of sustainability reporting, as it allows for the identification of issues that must be reported within the report.

In the context of double materiality, the Directive 2022/2464 identified two different perspectives: *impact materiality* (also known as Environmental and Social materiality) and *financial materiality*. The former specifies that sustainability-related issues are relevant if they give rise to impacts (negative, positive, actual, or potential) on people and/or the environment, and these are a consequence of the organization's direct operations and investments, its products, or services. This perspective involves a wide range of stakeholders (consumers, employees, communities, citizens, business partners). Impact materiality thus addresses the issues with an inside-out approach, considering all the impacts that the organization has on the environment externally to it. To quantify the damage of an actual impact, the "severity" parameter is used, which takes into account the severity and spread of the impact and how difficult it would be to remedy the damage; while for those with negative effects, the probability of the impact occurring is also added.

From a financial materiality perspective, on the other hand, sustainability issues relevant to the financial effects on the organization are put at the centre. It adopts an outside-in approach, where environmental and social issues are only significant when they financially impact the entity. Risk in this second approach is contextualized by considering the probability of occurrence and the extent it may have within the financial realm. Unlike the first perspective, this one represents an area of interest for the investor category, who are attentive to the organization's economic performance. The impact materiality and financial materiality perspectives can be thus combined into double materiality (EFRAG, 2023).

## *Sustainability Report*

Stakeholders are increasingly demanding detailed information from companies regarding their impact on the environment, society, and people. Therefore, it has become essential for companies to pay equal attention to both financial aspects and sustainability-related issues. This growing interest has prompted companies to adapt to stakeholders' demands by opting for alternative reporting formats. One of such formats is the sustainability report. This is a tool aimed at transparently disclosing all necessary sustainability information to stakeholders to assess the performance, impacts, risks, and opportunities of a company. Among the recipients of the document are all interested parties, including suppliers, customers, employees, local communities, investors, and financiers. This type of reporting offers a comprehensive overview of the company, analyzing aspects not included in the financial statement but which may influence future financial performance. Over time, several sustainability reporting frameworks have been developed, each with a given focus on specific aspects of ESG-related issues.

In the European Commission's 2001 Green Paper, the European Union defined the Sustainability Report as: "*The voluntary integration of social and environmental concerns by companies into their commercial operations and relationships with stakeholders*." (European Commission, 2001a, 2001b). This type of report should include both positive and negative impacts resulting from the company's activities. As mentioned in the sections above, the Directive 2014/95/EU, amending Directive 2013/34/EU, defined a set of rules regarding the disclosure of non-financial information, applicable only to large companies that are public-interest entities and have an average of 500 employees at the end of the financial year. Based on that directive, these companies are required to disclose information regarding several ESG-related aspects, including environmental, social, labor-related issues, diversity within senior management in terms of age, gender, respect for human rights, and anti-corruption. Global initiatives have focused on sustainability reporting, bringing about numerous changes through the introduction of new reporting requirements by laws, regulations, standards, and guidelines. Currently, the most used framework for drafting the sustainability report is the Global Reporting Initiative (GRI, see following section), a non-profit organization whose framework aims to define ESG performance reporting standards.

In brief, the structure of a sustainability report first includes a description of the organization and its purpose, providing an overview of the company itself, including vision, mission, strategy, structure, and governance. Secondly, it should be reported how the organization integrates sustainability into its business strategies and policies, providing an assessment of environmental impact by considering factors such as greenhouse gas emissions, social impact, including data on working conditions, human rights, diversity and inclusion, health and safety at work, and community involvement. Thirdly, economic data must be communicated, including information on economic value generation, tax payments, and investments. Additionally, the report should also indicate sustainability objectives that the company aims to achieve as well as the progress made towards achieving them, and a communication regarding the risks and opportunities related to sustainability issues should also be provided. Ultimately, it should indicate how an organization engages its stakeholders and meets their needs and expectations.

The sustainability report offers several advantages (Petrescu et al., 2020). Firstly, it demonstrates that business decisions are based on an understanding of environmental, social, and governance impacts, thus providing a foundation for the decision-making process. Furthermore, it provides transparency to stakeholders regarding business practices adopted to manage environmental, social, and governance impacts, contributing to building trust and strong relationships. It also serves as a useful tool for companies to identify potential areas for improvement, allowing them to evaluate their performance compared to others and to adopt any corrective measures. It also offers the opportunity to compare results achieved with those of competitors and industry peers, enabling an assessment of one's own positioning and identification of best practices to follow.

According to existing literature, some internal factors such as sector, and board structure, and other external factors such as legal, economic, cultural systems, can have a potential influence on the level of disclosure within a sustainability report (Karsam et al., 2024). For example, literature considering gender diversity within senior management highlighted women show greater interest in social issues and quality of life. Conversely, it has been shown that men are usually more oriented towards success and income.

Multiple studies have examined the impact of company size on disclosure quality and volume within a sustainability report and some scholars believe that sustainability reporting practices—being a complex phenomenon—cannot be framed or explained through a single theory. In this regard, Benvenuto et al. (2023) analyzed various theories, including legitimacy theory, stakeholder theory, institutional theory, voluntary disclosure theory, signaling theory, agency theory, political cost theory, and resource dependency theory to identify how sustainability reporting can be interpreted through those lenses. In brief, legitimacy theory is based on the concept of a social contract existing between a company and society. To maintain its operations in the long term, the organization must be seen as legitimate by other actors in the context in which it operates. Legitimacy can be achieved by acting in accordance with established norms. However, when an organization fails to meet the expectations of stakeholders, it is subjected to increased pressure—putting its legitimacy at risk—and therefore, it is necessary to voluntarily and selectively disclose information to reduce negative impact. Thus, in such circumstances, the development of sustainability reporting becomes extremely relevant to improving social perception. Stakeholder theory, related to legitimacy theory, suggests that organizations should create value for all stakeholders. As addressed in Chap. 1, this theory conflicts with the financial model, which focuses on creating value for shareholders. Therefore, it becomes essential for organizations to provide information regarding environmental, social, and economic impacts, as stakeholders can implicitly correct company behavior. Similarly, institutional theory combines the two aforementioned theories to understand how companies address social and institutional pressures. In this sense, sustainability reporting can be used as a tool to demonstrate that the organization operates within acceptable limits of activity. Further, voluntary disclosure theory suggests that companies, in order to reduce informational asymmetries, should be willing to voluntarily disclose expected information, but on the condition that the benefits outweigh the costs and that reputation improvement reduces the cost of capital. Therefore, according to this theory, organizations with better performance communicate clear and verifiable information, while companies with lower performance may provide more ambiguous and unverifiable information. Similarly, according to signaling theory, companies are predisposed to provide additional information in order to reduce informational asymmetries. This can also occur through the communication of sustainability initiatives undertaken. Agency theory, on the other hand, addresses the problem of the "conflict" between owner interests and manager interests, whose agency costs may increase due to such gaps. Therefore, even in this circumstance, sustainability reporting could help reduce the aforementioned asymmetry. According to political cost theory, companies decide to disclose sustainability-related data in order to reduce political costs, such as taxes, and to obtain benefits instead. Finally, resource dependency theory suggests that companies are not independent and autonomous, thus having to obtain resources within their context.

## GRI

The GRI has promoted a distinct reporting model for sustainability reporting based on a wide range of parameters related to the ESG dimensions. The GRI is a non-profit organization, based in Amsterdam, born with the aim of helping businesses and institutions provide "*a reasonable and balanced representation of an organization's positive and negative contributions to sustainable development*" (GRI, 2016). The GRI aims to promote a primary mechanism of accountability to ensure responsible behavior regarding the sustainability impacts of companies. The GRI has become one of the most widely adopted and shared frameworks for sustainability reporting, with over 15,000 organizations disclosing their ESG information in accordance with these standards as of 2021 (Abeysekera, 2022).

One of the main characteristics of GRI reporting is the numerous collaborations with international organizations, governments, and capital markets, including UNGC, OECD, and UN Working Group on Business & Human Rights, aimed at promoting sustainability development (GRI, n.d.-a, n.d.-b, n.d.-c). These collaborations contributed to the creation of broad general consensus around the Guidelines (or Standards)—and then to their subsequent global dissemination. Not surprisingly, the GRI reporting model is also characterized by the fact that all guidelines were created to address the widest possible range of stakeholders, including businesses, workers, civil society, investors, suppliers, academics, and sustainability reporting professionals. Furthermore, the establishment of the Global Sustainability Standards Board in 2014 and the related changes in the governance structure of the organization have strengthened the independence of GRI, which aims to achieve a certain degree of self-sufficiency, especially from a financial support perspective, now guaranteed by various sources of funding, including governments, businesses, foundations, partner organizations, and supporters.

Compared to the previous version (G4 Sustainability Reporting Guidelines), the main changes introduced by the latest GRI standards concern the structure and format of the information communicated: the model presents a modular and interconnected structure with the standards themselves, which are designed as a series to be followed to prepare a sustainability report. This new structure allows for the updating of individual standards or the addition of new ones without having to reconsider the entire document, thus simplifying reporting capacity and ensuring greater flexibility in relation to the evolution of different needs, contexts, and developments in sustainability reporting.

The GRI Standards are divided into four series, three of which are universal standards (Series 100) and three series of specific standards, one for each of the three dimensions of sustainability (GRI 200: economic issues, GRI 300: environmental issues, and GRI 400: social issues). The first three Universal Standards are applicable to all companies wishing to prepare a sustainability report "in accordance" with the GRI Standards, while Series 200, 300, 400, or Topic Specific Standards are selected in relation to the relevant themes for the specific company. For the first time, indeed, the GRI Standards provide the possibility of using only some standards or part of them to report specific information.

In detail, Series 100 includes the following Universal Standards:

- GRI 101: Reporting Principles (Foundation): These constitute the starting point where fundamental principles are introduced for defining the quality and content of the report, the requirements for using the different GRI Standards, and specific statements, or usage statements, required of organizations preparing a report in accordance with the GRI Standards.
- GRI 102: General Disclosure: Provides reporting of contextual information on the profile, strategy, ethics and integrity, governance, stakeholder engagement practices, and reporting process.

- GRI 103: Management Approach: It is useful for providing guidance on how the company manages material aspects that have a significant impact on stakeholders. It is designed to be used for each material issue, including those covered by specific GRI Standards. "Materiality" is indeed the principle that determines which issues are sufficiently relevant to warrant reporting. GRI 103, in fact, requires that a textual explanation be provided as to why the issue is material, where the impacts occur (the scope of an issue), and how the organization manages the impacts (GRI 101, 2016).

An organization intending to declare that it prepares a sustainability report in accordance with the GRI Standards must first apply the reporting principles of GRI 101. Reporting principles are divided into two groups: principles for defining the content of the report and principles for defining the quality of the report. The former helps to understand which content to include in the report; this involves an analysis of the activities and impacts of organizational activities and the real interests of stakeholders, as follows:

- Stakeholder inclusiveness: The organization must identify its stakeholders and define the process for their involvement.
- Sustainability context: The sustainability report must represent the organization's performance in the broader context of sustainability. This means that the organization's performance must be evaluated within the limits and demand for economic, social, or environmental resources at the sectoral, local, regional, or global level. It is equally important for a large company, for example, to report on both global impacts such as climate change and local factors such as community development.
- Materiality: The sustainability report must cover issues that reflect the economic, social, and environmental impacts (positive or negative) associated with the company's activities or that influence the decisions and assessments of its stakeholders substantially.
- Completeness: The contents of the report must include the list of material issues analyzed, the perimeter of the issues (topic boundaries) in a manner sufficient to reflect significant economic, social, and environmental impacts and to enable stakeholders to assess the company's performance in the specific reporting period.

The reporting principles for defining the quality of the sustainability report help, instead, ensuring the quality and proper representation of the information included in the report. The quality of information is an important factor in enabling stakeholders to make realistic and reasonable assessments and to take appropriate actions. The principles dedicated to defining quality are as follows:

- Comparability: The company must report information consistently, in order to allow stakeholders to analyze performance trends over time (compared to past performance or set goals) or to compare them with those of other companies. To facilitate comparability over time, consistency must be maintained in the use of calculation methods employed and in the presentation of information (e.g., page layout), explaining methodologies, estimates, and assumptions used in data processing and any updates over time. Furthermore, the company must include both absolute data (total impact values, e.g. tons of waste) and indexed data (i.e., normalized, e.g. amount of waste per unit of production).

- Balance: It requires that data should reflect both positive and negative aspects of performance in such a way as to provide stakeholders with an impartial view for a weighted evaluation of overall performance. Choices, omissions, or formats that could unduly influence decisions or judgments must be avoided, clearly distinguishing between objective data and data subject to interpretation.
- Reliability: The company must collect, record, compile, analyze, and present information and processes used in the preparation of the report in such a way that they are examinable and their quality and materiality can be defined. Third parties other than those who prepared the report must be able to determine the truthfulness of the contents of the sustainability report and the supporting documentation. Information unsupported by concrete evidence cannot be included unless it represents material information, and the report provides unequivocal explanations of uncertainties related to such information. In these cases, the company must report recalculated historical data to make them comparable or, if recalculation is not possible, must provide sufficient explanations for interpreting current information.
- Accuracy: The information (qualitative or quantitative) subject to reporting must be sufficiently accurate and detailed so as to enable stakeholders to assess the performance of the organization. The characteristics that define accuracy vary depending on the nature of the information and the type of user. For example, qualitative information may be influenced by the degree of clarity and detail or the scope of the fixed topic, while quantitative information may depend on the methods used to gather information
- Clarity: Data and information must be presented in such a way as to be understandable, and accessible to all stakeholders, and accessible in paper format or through other channels. It is important for stakeholders to be able to easily access information and data, which can also be presented in the form of graphs and tables.

Series GRI 200, 300, 400 include specific standards that can be used to report information on an organization's impacts related to ESG issues (e.g., direct economic impacts, water, or employment; GRI, n.d.-a). Thanks to the modular structure, frequent updating of these Standards is possible. For example, in 2018, GRI 303-Water, GRI 403-Health and Safety standards were updated, and GRI 207-Tax was introduced. In practice, companies often use the so-called sector supplements or GRI Sector Disclosures that integrate specific guidelines and indicators depending on the reference sector.

Sector Standards represent the most substantial and significant change introduced by the 2021 revision; these have been added alongside the existing Universal and Topic Standards with the aim of improving the quality, completeness, and consistency of reports from organizations operating in the same sector (GRI, n.d.-b). The Standards included in this category provide an overview of sector characteristics and list the material topics and relevant information typically disclosed by organizations in a specific sector. Furthermore, the adoption of these standards is mandatory for organizations belonging to regulated sectors. This aims to make reports comparable as quickly as possible. The sectors already regulated include the Oil and Gas Sector (GRI 11), Coal Sector (GRI 12), and Agricultural, Aquacultural, and Fishing Sector (GRI 13), and overall, more than 40 sectors will be regulated.

## *Integrated Report*

Criticisms have emerged from scholars who believe that the sustainability report presents a major limitation in that it does not integrate financial and non-financial information (Van Wyk & Els, 2023). Consequently, integrated reporting has been proposed to address this issue through the inclusion, in a single document, of both quantitative and qualitative information.

The International Integrated Reporting Council (IIRC) was established in 2010 at the proposal of two organizations, both engaged in sustainability: the Prince's Accounting for Sustainability Project and the GRI. The IIRC included 40 members including heads of the IASB, FASB, International Federation of Accountants, and International Organization of Securities Commissions, the leaders of the four largest professional services networks in the world, namely, Deloitte, EY, KPMG, and PwC, the heads of major British accounting bodies, and the financial directors of major multinationals such as Nestlé, Tata, and HSBC. The founders shared the opinion that communication about value creation or diminution constituted a next step in the evolution of corporate reporting. To address this need, the IIRC developed the International <IR> Framework (International Integrated Reporting Council, 2020). In 2014, in order to involve other standard-setting entities, the IIRC implemented an initiative called the Corporate Reporting Dialogue (IFRS Foundation, 2024a). The purpose of the IIRC was to promote dialogue and alignment between financial reporting standards (IASB and FASB) and sustainability reporting standards (GRI and SASB). However, GRI representatives expressed dissatisfaction with the proposals, arguing that the Corporate Reporting Dialogue had been turned into a tool for promoting <IR>, thus raising significant concerns about GRI's participation in the Corporate Reporting Dialogue.

In spite of these efforts, issues regarding the positioning in corporate reporting remained unresolved due to a lack of clarity on the definition of integrated reporting and its placement. This prompted the IIRC to focus its attention on the integrated reporting processes, supporting the so-called integrated thinking (IFRS Foundation, 2024b). In November 2020, the idea of merging the IIRC and SASB into the Value Reporting Foundation was announced. In this regard, O'Dwyer et al. (2024) report an interview conducted with a former CEO, who described the merger as a strategic union to consolidate their position in various jurisdictions and to remain relevant to developments within the IFRS Foundation. Subsequently, the Value Reporting

Foundation requested greater alignment between financial accounting and sustainability standards, leading the IFRS Foundation to issue a draft concerning the International Sustainability Standards Board (ISSB).

The key concept underlying the IIRC is one to integrate the traditional balance sheet, management report, governance and remuneration report, and sustainability reports, in order to highlight the most relevant features of these documents and their interactions and how they influence an organization's ability to create value (O'Dwyer et al., 2024).

According to the IFRS Foundation's statement, the main objective of integrated reporting is to improve the quality of available information, promote an alternating reporting approach capable of integrating different aspects of reporting while also communicating how these aspects affect the organization. Furthermore, this approach aims to enhance accountability and resource (or capital) management—financial, intellectual, social, and human—while promoting a better understanding of their dependencies. Integrated reporting is designed to highlight the impact these capitals have on the business.

In a 2011 Discussion Paper, the voluntary nature of integrated reporting was evident, as no key performance indicators or measurement or reporting methods are outlined. Therefore, organizations must identify both material aspects and how to communicate them (Integrated Reporting, 2011). In this regard, a critique has been proposed by Flower (2015), who emphasized that the decision to make this report voluntary represents a significant regression, as it loses the qualification of the main balance sheet, becoming simply an additional document.

For an integrated report to comply with the <IR> Framework, some minimum requirements must be met to allow a sufficient level of comparison between different organizations through the implementation of these principles. The purpose of the Framework is to define Guiding Principles and Content Elements outlining the information to be reported. Although primarily aimed at the private sector and for-profit companies of any size, it can also be adapted by public sector enterprises and non-profit organizations. The integrated report must be based on seven Guiding Principles and eight Content Elements. Regarding the former, it must be able to provide a clear view of the strategy and future prospects, explaining at the same time how the strategy contributes to value creation. Furthermore, the report should highlight the connection between different factors and relationships with key stakeholders and how the organization meets their needs. Moreover, the report should focus on relevant issues (materiality) that generate or may generate significant impacts on the organization's ability to create value. In this regard, a report must be clear, reliable, and complete, including both positive and negative aspects. Finally, through a coherent presentation of information, a report should allow comparison with other organizations. The Content Elements include an introduction to the organization and its external context, outlining the organization's activities. Additionally, they examine the governance structure and how it contributes to value creation and provide a description of the operating model, also analyzing risks and opportunities. The report also includes strategy planning and resource management, defining the objectives and how an organization intends to achieve them. Past and current performance are also assessed, subsequently identifying future challenges. Finally, the criteria used by the organization to determine which aspects to report in the integrated report and how these are evaluated are discussed.

Since 2010, only South Africa has made <IR> mandatory for listed companies (Wachira et al., 2020). In this regard, the idea of mandating integrated reporting in South Africa originated in 1994, with the introduction of the first King Code of Corporate Governance Principles, known as "King I," in honor of Mervyn King, developed as South Africa was transitioning to democracy. Subsequently, in 2002—in the aftermath of the Johannesburg Summit—King II' code was developed, which introduced the concept of "Integrated Sustainability Reporting." This document was also developed based on the Global Reporting Initiative, and some parts of it were adopted by the New York Stock Exchange, thus becoming one of the major influences on corporate governance principles internationally.

Whether and how integrated reporting can promote sustainability goals is an aspect on which academics and professionals seem to have diverging views. Indeed, a debate about the benefits and usefulness of integrated reporting is currently ongoing, especially following the issuance of the European Directive on non-financial reporting, which required large companies that are public-interest entities to include information on their environmental, social, human rights, and anti-corruption impacts. According to Barth et al. (2017), integrated reporting is on the cusp of a global adoption phase and over time, it is expected to establish itself as the primary standard for corporate reporting. Nowadays, however, despite IR being adopted internationally by a very broad network of organizations from different sectors, the Framework has not yet achieved the IIRC's mission, namely to establish integrated reporting and thinking within mainstream business practice as the norm in the public and private sectors. Along similar lines, according to Busco (2018), integrated reporting has the potential to become the most widely spread reporting practice as it expands the scope of traditional accounting and reporting systems, by using a more connected approach of how financial and non-financial sources of capital that a business use can be strategic to the achievement of long-term sustainable value creation. This view has been supported by the United Nations Global Compact, who stated that the level of information integration due to the adoption of integrated thinking and reporting encourages and supports the integration of sustainability in strategic planning, decision-making, and operations (Adams, 2015). Compared to traditional social reporting, the application of integrated thinking or principles such as connectivity of information allows overcoming the limitation of non-connection with financial data. This aspect contributes to improving the quality and reliability of the information transmitted to financial capital providers, who are the main recipients of the integrated report. Thanks to the large amount of managed information and its interconnections, the integrated report allows the company to have a holistic view of how the company creates, modifies, or destroys value in the short, medium, and long term, also facilitating the identification of emerging risks and opportunities.

Beyond the actual likelihood that the organization will achieve the aforementioned benefits through the adoption of integrated reporting, which is still to be demonstrated, some criticisms and opposing positions towards this new reporting tool have recently emerged. A criticism levelled at the integrated report concerns how it addresses the issue of sustainability. Stubbs and Higgins (2018) believe that the integrated report can be considered as an incremental first phase of sustainability reporting, rather than a revolutionary transformation of current financial and sustainability reporting approaches. The two authors argue that the integrated report is in its early adoption phase and could lead to numerous innovations in disclosure mechanisms, although currently, various integrated reports represent a sort of transition from previous sustainability reports rather than a radical innovation, mainly due to the lack of non-financial standards that makes it difficult to compare companies' performances and may raise doubts and uncertainties about the reliability, completeness, and coherence of the information presented in the integrated report. Another pertains to the fact that users of the IR Framework must constantly select the information that the report should include, choosing what is considered noteworthy (material information) and excluding the rest (non-material information). Inevitably, this choice combines the various legitimate interests of stakeholders and requires a compromise that is sometimes difficult to find and that could only favor investors rather than the broader category of stakeholders. For example, Flower (2015) is critical of the IIRC's concept of value because it is "value for investors" and not "value for society." In line with this view, Brown and Dillard (2014) argue that the integrated report, as conceived by the IIRC, provides a very limited and one-sided approach to assessing and reporting on sustainability issues.

## *SMEs*

Small and medium-sized enterprises (SMEs) are defined by the European Union according to the following classification:

- Micro enterprises, which employ fewer than 10 people and have an annual turnover and balance sheet of less than two million euros.
- Small enterprises, which employ fewer than 50 people and have an annual turnover and balance sheet below ten million euros.
- Medium enterprises, which employ fewer than 250 people and have an annual turnover of less than 50 million euros and a balance sheet below 43 million euros.

The European Union comprises over 23 million SMEs, representing approximately 99% of all businesses operating within the EU.

This extensive network of enterprises accounts for two-thirds of all private sector jobs, employing around 76 million people and representing 49% of the total employment (Eurostat, 2023). Additionally, SMEs generate over half of the total added value created by companies in the EU, thus being the key driver of the European economy. From this, it can be argued that the role of SMEs within the European area is critical for the realization of the 2030 Agenda, considering that SMEs contribute up to 70% of global pollution (Hillary, 2000).

SMEs in Europe's health care sector are pivotal for driving innovation, job creation, and tackling critical healthcare challenges through their contributions to research, development, and the commercialization of new medical products. While the sector primarily consists of non-market output, indicating significant public provision of healthcare services across most EU countries, large enterprises, though fewer in number, employ a significant portion of the business labor force and generate a substantial share of the value added (Gutiérrez-Hernández & Abásolo-Alessón, 2021). In the European medical technology sector which is highly innovative, SMEs constitute 80% and are vital for job creation, particularly in research and manufacturing. This sector employs just over half a million people across Europe, underscoring the importance of SMEs in providing highly skilled jobs and fostering innovation within the healthcare sector (MedTech Europe, 2021). Further, biopharmaceutical SMEs play a key role in the discovery and development of new medicines, especially in sectors facing technological challenges or with lower expected market potential. These enterprises are essential for addressing areas such as biological approaches, potential pandemics, infectious diseases, and rare diseases (European Federation of Pharmaceutical Industries and Associations, 2016).

Research indicates that environmental and social concerns are increasingly important for SMEs (Revell et al., 2010). However, due to their distinctive characteristics, SMEs often face greater challenges in implementing ESG practices compared to larger firms. Indeed, scholars (Shalhoob & Hussainey, 2023; Gholami et al., 2022) have focused on the unique challenges that SMEs face when it comes to implementing ESG practices. From this literature, the main barriers explaining the reasons SMEs reticence in adopting ESG can be broadly categorized into external and internal factors (Johnson & Schaltegger, 2016).

## External Factors

External deficiencies faced by SMEs encompass inadequate external incentives, complexity of internationally designed standards, inadequacy of management tools, and heterogeneity of SMEs. Governmental regulations and market pressures are perceived as significant barriers for SMEs contemplating involvement in ESG practices (Hillary, 2004; Seidel et al., 2009). Historically, the limited regulatory requirements regarding sustainability disclosure for SMEs, along with minimal customer pressure, led SMEs to perceive ESG activities as not worth pursuing (Johnson & Schaltegger, 2016). Existing ESG disclosure frameworks are designed based on the data collection capabilities and standardized structures typical of large corporations and this dissuades SMEs from embarking on the sustainability journey as they are often incompatible with the flexible and dynamic business model of SMEs (Setyaningsih et al., 2023). Furthermore, ESG performance metrics tend to prioritize qualitative measures over financial ones, making it challenging for SMEs to gather relevant data and resulting in increased costs due to the lack of specialized expertise required to assess and report on ESG indicators (Ozkan et al., 2023). In addition, while interest in sustainability disclosure has grown in recent years, the varied nature of SMEs in terms of customers, business models, and organizational structures makes it challenging to disclose comparable data among SMEs with different conceptual frameworks.

### Internal Factors

Within the internal factors it is possible to identify the following elements: lack of human and financial resources, lack of awareness, lack of knowledge and expertise, lack of perceived benefits, and proprietary information risks.

Employees in SMEs often have diverse responsibilities within the organization, which can limit the time available for specific ESG training (Burke & Gaughran, 2007). Due to the technical knowledge and skills required for ESG disclosure, as well as the need for organizational planning and managerial competencies, SMEs may find themselves at a disadvantage compared to larger businesses. Furthermore, despite the wealth of information and resources available to SMEs, including exposure to market influences, opportunities, and threats, they often struggle to effectively adapt to current trends and utilize this valuable information to respond to pressures within their networks (Brodny & Tutak, 2022). In addition, SMEs' owners and managers often fail to recognize the financial and non-financial benefits that ESG practices can bring to their businesses (Johnson & Schaltegger, 2016). The advantages resulting from ESG investments are not immediately apparent; they often become visible only after some time, which may discourage SMEs from investing in such activities due to their short-term focus resulting from limited resources (Biondi et al., 2002).

Owners and managers of SMEs have often limited awareness of the environmental and social impact of their business activities (Johnson & Schaltegger, 2016; Seidel et al., 2009). Even when owners and managers recognize the impact of their activities and the potential benefits of implementing ESG practices, they may lack the knowledge to address these issues effectively (Seidel et al., 2009). Additionally, SMEs typically do not take action to improve their environmental footprint as they generally perceive themselves as unaffected by sustainability issues (Revell & Blackburn, 2007) and consider their impact on the environment and society to be minimal due to their small size compared to the overall economy (Lawrence et al., 2006). However, considering the cumulative impact of all SMEs, such behavior contradicts the principle of sustainable development, given that SMEs have a significant impact on the European economy compared to larger firms.

The comparatively limited resources available to SMEs are the primary reason for the low implementation of ESG practices (Santos et al., 2022). Larger enterprises have greater financial resources compared to SMEs, enabling them to invest more easily in ESG initiatives, take on more social responsibility and, build stronger brands. Additionally, compliance with non-financial information disclosures can be costly, particularly for SMEs, hence carrying a negative effect of profitability at least in the short term (Chen et al., 2021). These costs involve the collection and preparation of data and the potential loss of competitive advantage due to the exposure of proprietary information to competitors. In fact, ESG disclosure may reduce SMEs' competitive advantage, given their lower market diversification and focus on niche markets (Gjergji et al., 2021).

## Opportunities

SMEs embarking on the sustainability path can potentially benefit from various financial opportunities. These include a reduction in the cost of debt, the availability of tax incentives and government subsidies, mitigation of ESG-related risks, long-term cost savings, and opportunities for market expansion. Financing institutions, such as banks, often impose various types of internal controls on SMEs to address the knowledge gap resulting from information asymmetries between the two parties. As a result, sustainability disclosures become valuable sources of information for banks and other financing stakeholders, enabling SMEs to access lower interest rates. This is because increased transparency and the risk mitigation capacity provided by ESG reporting can lead to lower financial risks for SMEs (Kim & Li, 2021; Singh & Pillai, 2022). Indeed, research suggests a negative correlation between sustainable performance and financial outcomes (Chen et al., 2023), implying that companies with consistently high ratings for voluntary ESG disclosure tend to pay lower interest rates when seeking debt financing. However, some studies have reported conflicting findings regarding the effect of ESG performance on the cost of debt, showing no statistically significant positive impact in some instances (Apergis et al., 2022). To this respect, the publication and disclosure quality of ESG reports can influence the creditworthiness of SMEs as part of their financial performance (Kim & Li, 2021). Disclosure quality plays a crucial role in addressing information asymmetries between owners and managers, as stakeholders are interested in both qualitative and quantitative data (Venturelli et al., 2019).

Governments worldwide have taken steps towards encouraging SMEs to align with sustainability goals by implementing tax incentives and subsidies (Picas et al., 2021). Nonetheless, to date, a comprehensive sustainable tax system has not yet been established. Remaining within a financial perspective, as with large companies, SMEs can achieve cost savings by implementing sustainable practices such as improving energy efficiency and reducing waste disposal. A recent study by Henisz et al. (2019) revealed that the reduction in operating expenses resulting from efficient resource utilization could impact up to 60% of a company's operating profit. Stakeholders, including investors, view efficient resource use as a reliable indicator of management's approach to resource allocation. Finally, through the implementation of ESG processes, companies may be able to satisfy new customers' needs as well as proposing products and services that were previously unavailable within the market (Seidel et al., 2009). In terms of non-financial opportunities, implementing ESG practices may increase talent attraction, retention, and motivation, as well as firms' reputation, professional image, and brand value (Kotsantonis et al., 2016). In fact, if greenwashing has a negative effect on the image of the company, transparent and reliable ESG reporting can act as a powerful branding and marketing tool (Seidel et al., 2009).

### ESG Reporting

Despite their relevance within the European context, SMEs have been mostly overlooked with respect to the sustainable development agenda (Setyaningsih et al., 2023). While SMEs are not mandated to report under the CSRD unless listed on regulated markets, they can voluntarily adopt simplified standards which are tailored to their capabilities, for easier comparison and verification making reporting more manageable for stakeholders such as banks, clients, and investors. The disclosed data will undergo limited third-party assurance, where an auditor evaluates the information to ensure accuracy and reliability.

At the time of writing, a public consultation on new Sustainability Reporting Standards for SMEs under the CSRD is underway (European Commission, 2024). Indeed, EFRAG has released two Exposure Drafts regarding sustainability reporting standards for SMEs, inviting stakeholders to provide feedback until May 21, 2024. EMAS stakeholders are encouraged to contribute their insights and feedback during this period. While the current European Sustainability Reporting Standards cater to large companies, EFRAG is currently developing proportionate standards for listed SMEs and voluntary standards for non-listed SMEs (VSMEs, EFRAG, 2024). The ESRS LSME exposure draft targets SMEs classified as public-interest entities. This iteration of the ESRS is structured as a delegated act, slated to take effect on January 1, 2026, with an optional two-year opt-out period. Its objective is to offer reporting criteria adjusted to the size and intricacy of the operations of LSMEs, aiding them in enhancing access to financing and furnishing standardized sustainability data. The VSME represents a voluntary sustainability reporting standard tailored for non-listed SMEs (VSME). According to EFRAG, although these standards are not mandatory, the aim is to aid SMEs in providing sustainability data when requested by business partners and to facilitate SMEs' participation in the transition towards a sustainable economy. The public consultation offers an opportunity to improve alignment with other reporting schemes in the two exposure drafts for sustainability reporting standards for SMEs. Both the mandatory ESRS LSME ED and the voluntary VSME ED are expected to become significant reporting standards.

## ESG Providers and Measures

The disclosure of information within corporate reports forms the basis for ESG scores and ratings, providing stakeholders with comprehensive insights into companies' ESG practices. These reports serve as a primary source of data for ESG scoring systems and rating agencies, enabling the quantification and comparative evaluation of companies' sustainability performance over time (Roncalli, 2024). Indeed, several providers of ESG data and scores exist nowadays, following a progressive concentration of the industry and a period of consolidation in the 2010s. If, at the beginning of the 2000s, around three quarters of the total number of sustainable research organizations were located in Europe, most of which being founded in the 1990s including Global Engagement Services in Sweden,[1] Oekom Research AG in Germany,[2] Sustainable Asset Management Ltd. in Switzerland,[3] Vigeo in France,[4] and Dutch Sustainability Research in the Netherlands,[5] today the industry of extra-financial analysis and ESG ratings is dominated by a few big players.

These are ISS-Oekom, MSCI, Refinitiv, Reprisk, S&P Global, Sustainalytics and Moody's, with the market being therefore dominated by large Anglo-Saxon companies (US and UK; ORSE, 2007). According to a recent classification (Demartini, 2020), four types of entities characterize the ESG provider industry:

- Market data companies which include financial information providers, index sponsors, and stock exchanges
- Financial rating agencies
- Specialized ESG companies
- Technology start-up firms which predominantly use artificial intelligence, big data, natural language processing methods

These entities differ in the scope of activities and services which comprise: provision of raw data, provision of processed data (indicators, scores, and ratings), production of ESG indexes, ESG screening, normative analysis, engagement monitoring, portfolio analysis, ESG controversy tracking, proxy advisory services, and consultancy.

The business model of extra-financial rating agencies is distinct from that of credit rating agencies, being primarily based on the investor-pays principle rather than the issuer-pays principle traditionally employed by credit rating agencies (Bush, 2022). However, another major difference exists. Credit rating agencies operating within the European Union fall under the supervision of the European Securities and Markets Authority, in accordance with Regulation 462/2013/EU and Directive 2013/14/EU. In the United States, oversight of these agencies, which needs to be registered as nationally recognized statistical rating organizations, is

[1] Caring Capitalism was founded in 1992 and renamed Global Engagement Services in 2003.

[2] The environmental publishing company ökom was established in 1989. In 1993, ökom GmbH was established to conduct environmental research. Oekom research AG gained independence in 1999 and shifted its focus to corporate responsibility assessments.

[3] SAM was established in 1995. In 1999, SAM partnered with Dow Jones to introduce the Dow Jones Sustainability Indexes. In 2001, SAM launched the inaugural global sustainable water fund, known as the SAM Sustainable Water Fund. In 2006, Robeco acquired SAM, and in 2013, it was rebranded as RobecoSAM.

[4] Established in 1997, Arese was France's pioneering SRI rating agency. By June 2002, it underwent a transformation, rebranding as Vigeo and coming under the leadership of Nicole Notat, the former secretary general of the CFDT labor union. Subsequently, in June 2005, Vigeo merged with the Belgian agency Ethibel.

[5] DSR originally served as the research division of Triodos Bank, a Dutch institution specializing in sustainable finance, established in 1980. In 2008, it underwent a name change, rebranding itself as Sustainalytics.

facilitated by the Office of Credit Ratings which collaborates with the Securities and Exchange Commission. In fact, with the anticipated growth of the ESG ratings market, regulatory supervision and investor protection have become increasingly significant. For instance, in April 2022, the European Commission initiated a focused consultation regarding the functionality of the ESG rating market within the EU and the incorporation of ESG factors in credit ratings (European Commission, 2022). The summary report, comprising insights from 204 responses, was published in August 2022, with the following findings: "*The large majority of respondents (over 84%) consider that the market is not functioning well today. On the quality of ESG ratings, two thirds of respondents consider the quality to be fine to very good, with about one third considering it poor. A large majority of respondents (83%) consider that the lack of transparency on the methodologies used by the providers is a problem in the ESG ratings market. The vast majority of respondents (91%) also consider that there are significant biases with the methodology used by providers [...] Almost all respondents (94%) consider that intervention is necessary, of which the large majority (80%+) support a legislative intervention with the remainder supporting the development of non-regulatory intervention in the form of guidelines, code of conduct. Respondents largely indicated (90%+) that the main element to be addressed by the intervention should be improving transparency on the methodology used by ESG rating provider [...] The vast majority of respondents (82%) consider that ESG rating providers should be subject to some form of authorization/registration regime in order to offer their services in the EU.*" (EC, 2022b, pages 3–4). These findings corroborate earlier analyses that identified the primary deficiencies in ESG ratings, including insufficient coverage, data quality issues, and a lack of transparency regarding methodologies employed by ESG rating providers (Boffo & Patalano, 2020). Consequently, it is reasonable to expect that regulatory authorities will implement measures to address these concerns and introduce regulatory safeguards for the utilization of ESG ratings in the foreseeable future.

Among the top players in the ESG market, the following four providers represent a large section of the offer currently available:

- MSCI, Morgan Stanley Capital Investment, formerly a subsidiary of the Morgan Stanley group, creates market indices for capital management companies. In 2010, MSCI acquired RiskMetric, one of the first providers of sustainability data in the United States, and launched its MSCI ESG Research rating, which now evaluates 9000 companies worldwide. It analyzes 37 key themes (including pollution, waste production, human capital, and institutional behaviors) among the data available on government portals and in corporate communications and combines them to issue a rating on a scale ranging from CCC to AAA.
- Sustainalytics. Publishes ESG ratings for 13,000 companies worldwide, but its main task is to evaluate funds created by its parent company, Morningstar. It offers a percentage score (from 0 to 100) by analyzing 70 indicators divided into three themes: environment, society, and governance. For each indicator, it evaluates readiness (how well management systems and corporate policies can handle ESG risks), disclosure (how transparent the company is in communications

regarding ESG issues), and performance (based on the analysis of ESG incidents it has faced).
- ISS, Institutional Shareholder Services, with its ISS ESG rating, evaluates 12,000 issuers and 25,000 funds. Controlled by Deutsche Börse, the company that owns the Frankfurt Stock Exchange, it measures the environmental and social impact of companies, providing a mark on a 1-point scale from D– to A+.
- Refinitiv and FTSE Russell. Both controlled by the London Stock Exchange Group, they publish ESG ratings for around 13,000 issuers. Refinitiv analyzes transparency and commitment to ESG issues, evaluating emissions, resource usage, and respect for human rights, for example. FTSE Russell provides support to corporate administrators by evaluating not only the virtuousness and proper communication of these non-financial information but also exposure to ESG risks, with over 200 indicators.

However, a lack of ESG data standardization and consequent discrepancies in ESG ratings presents a major issue. Indeed, a famous illustration of this issue has been proposed by Berg et al. (2019) who examined the divergence in ESG ratings provided by six leading agencies: KLD, Moody's ESG, MSCI, Refinitiv, S&P Global, and Sustainalytics. Utilizing the ESG metrics offered by these providers, the researchers created synthetic ratings by applying a standardized framework consisting of 64 indicators. Based on their analysis, the divergence observed in ESG ratings stems from three primary sources, that is, measurement (rating agencies employ different ESG metrics to evaluate the same indicator); scope (ratings are derived from dissimilar sets of ESG indicators); weight (varying levels of relative importance are assigned to different ESG indicators). Their research reveals that measurement constitutes 56% of the overall disparity, with scope contributing 38% and weight divergence 6%. Since the publication of that study, ongoing discussions among both practitioners and academics have centered on standardization issues regarding data and methodologies in ESG rating systems (Billio et al., 2021). These discussions underscore the importance of addressing standardization issues to enhance the reliability and comparability of ESG ratings across the industry.

## *Refinitiv Platform*

Focusing on one of the major data providers - from which ESG and company data were sourced for the analyses later presented in Chaps. 4, 5, and 6 - Refinitiv's ESG scores are designed to measure a company's ESG performance, commitment, and effectiveness. The ESG score in Refinitiv covers 10 main themes, including emissions, product innovation with a focus on environmental protection, human rights, shareholder diversity, and so forth. Refinitiv also provides a comprehensive ESG score, which considers any relevant ESG issues the company may have faced. The methodology for assigning Refinitiv's ESG score is based on the following calculation principles:

- Unique weightings for ESG materiality assessment: The importance of ESG factors varies among companies in different sectors. The materiality score, ranging from 1 to 10, varies based on the type of metric used and is related to the company's business activity. Lack of immaterial information does not penalize the ESG score. Conversely, the lack of material information will negatively impact the ESG score.
- Actions implemented by companies regarding their commitments are verified to amplify the impact of issues on the overall ESG score. The scoring methodology also considers the market capitalization bias suffered by large companies, introducing severity weights to ensure that scores assigned to any issues that may arise in a corporate context are adjusted based on the company's size.
- Application of sectoral and national benchmarks to facilitate comparability within similar groups of companies.
- Application of a methodology that can assign a score based on percentile rank, thus producing a score ranging from 0 to 100, in addition to providing further easily understandable information.

The Refinitiv platform employs over 700 research analysts specialized in content analysis (Erlingsson & Brysiewicz, 2017) to collect ESG data, with different language skills. The score results from over 630 measures manually processed by analysts for each company. Each measure undergoes a process of information standardization to ensure comparability across all companies in the sector. The database is constantly updated, and data is refreshed weekly, including the recalculation of ESG scores. Such updates may include adding a new company to the database, updating the latest fiscal year, or including new potential issues arising in the corporate context. In most cases, reported ESG data is updated once a year in line with the companies' own ESG reporting disclosure. News and any ESG-related issues are continuously updated as these events occur and are detected by global media.

## References

Abeysekera, I. (2022). A framework for sustainability reporting. *Sustainability Accounting, Management and Policy Journal, 13*(6), 1386–1409. https://doi.org/10.1108/SAMPJ-08-2021-0316

Adams, C. A. (2015). The international integrated reporting council: A call to action. *Critical Perspectives on Accounting, 27*, 23–28. https://doi.org/10.1016/j.cpa.2014.07.001

Apergis, N., Poufinas, T., & Antonopoulos, A. (2022). ESG scores and cost of debt. *Energy Economics, 112*, 106186. https://doi.org/10.1016/j.eneco.2022.106186

Barth, M. E., Cahan, S. F., Chen, L., & Venter, E. R. (2017). The economic consequences associated with integrated report quality: Capital market and real effects. *Accounting, Organizations and Society, 62*, 43–64. https://doi.org/10.1016/j.aos.2017.08.005

Bennett, M., & James, P. (1999). Key themes in environmental, social and sustainability performance evaluation and reporting. In M. Bennett, P. James, & L. Klinkers (Eds.), *Sustainable measures: Evaluation and reporting of environmental and social performance* (pp. 29–74). GreenLeaf Publishing.

Benvenuto, M., Aufiero, C., & Viola, C. (2023). A systematic literature review on the determinants of sustainability reporting systems. *Heliyon, 9*(4), e14893. https://doi.org/10.1016/j.heliyon.2023.e14893

Berg, F., Kölbel, J., & Rigobon, R. (2019). Aggregate confusion: The divergence of ESG ratings. *Review of Finance*. https://doi.org/10.2139/ssrn.3438533

Billio, M., Costola, M., Hristova, I., Latino, C., & Pelizzon, L. (2021). Inside the ESG ratings: (Dis)agreement and performance. *Corporate Social Responsibility and Environmental Management, 28*(5), 1426–1445. https://doi.org/10.1002/csr.2177

Biondi, V., Iraldo, F., & Meredith, S. (2002). Achieving sustainability through environmental innovation: The role of SMEs. *International Journal of Technology Management*. Advance online publication. https://doi.org/10.1504/IJTM.2002.003074.

Boffo, R., & Patalano, R. (2020). *ESG investing: Practices, progress and challenges. OECD Paris*. Accessed March 11, 2024, from www.oecd.org/finance/ESG-Investing-Practices-Progress-and-Challenges.pdf

Brodny, J., & Tutak, M. (2022). Digitalization of small and medium-sized enterprises and economic growth: Evidence for the EU-27 countries. *Journal of Open Innovation: Technology, Market, and Complexity, 8*(2), 67. https://doi.org/10.3390/joitmc8020067

Brown, J., & Dillard, J. (2014). Integrated reporting: On the need for broadening out and opening up. *Accounting, Auditing & Accountability Journal, 27*(7), 1120–1156. https://doi.org/10.1108/AAAJ-04-2013-1313

Buhr, N. (2007). Histories of and rationales for sustainability reporting. In J. Unerman, J. Bebbington, & B. O'Dwyer (Eds.), *Sustainability accounting and accountability* (pp. 57–69). Routledge.

Burke, S., & Gaughran, W. F. (2007). Developing a framework for sustainability management in engineering SMEs. *Robotics and Computer-Integrated Manufacturing, 23*(6), 696–703. https://doi.org/10.1016/j.rcim.2007.02.001

Busco, C. (2018). Make sustainable development goals happen through integrated thinking: An introduction. In *Sustainable development goals and integrated reporting* (1st ed., pp 19). Routledge. Retrieved from eBook ISBN 9780429027314.

Bush, C. (2022). Dealing with the conflicts of interest of credit rating agencies: A balanced cure for the disease. *Capital Markets Law Journal, 17*(3), 334–364. https://doi.org/10.1093/cmlj/kmac012

Carroll, A. B. (2008). A history of corporate social responsibility: Concepts and practices. In A. M. Andrew Crane, D. Matten, J. Moon, & D. Siegel (Eds.), *The Oxford handbook of corporate social responsibility* (pp. 19–46). Oxford University Press.

Chen, L., Yuan, T., Cebula, R. J., Shuangjin, W., & Foley, M. (2021). Fulfillment of ESG responsibilities and firm performance: A zero-sum game or mutually beneficial. *Sustainability, 13*(19), 10954. https://doi.org/10.3390/su131910954

Chen, S., Song, Y., & Gao, P. (2023). Environmental, social, and governance (ESG) performance and financial outcomes: Analyzing the impact of ESG on financial performance. *Journal of Environmental Management, 345*, 118829. https://doi.org/10.1016/j.jenvman.2023.118829

Demartini, A. (2020). *Provision of non-financial data: Mapping of stakeholders, products and services. AMF risks and trend mapping, December*. Accessed March 12, 2024, from https://www.amf-france.org/sites/institutionnel/files/private/2020-12/mapping-esg-publication.pdf

Edgley, C., Jones, M. J., & Atkins, J. (2015). The adoption of the materiality concept in social and environmental reporting assurance: A field study approach. *The British Accounting Review, 47*(1), 1–18. https://doi.org/10.1016/j.bar.2014.11.001

EFRAG (European Financial Reporting Advisory Group). (2023). *Implementation guidance for the materiality assessment*. Accessed March 18, 2024, from https://www.efrag.org/Assets/Download?assetUrl=%2Fsites%2Fwebpublishing%2FMeeting%20Documents%2F2302241032237237%2F03-02%20Materiality%20Assessment%20Implementation%20guidance%20clean%20SRB%20231025.pdf

EFRAG (European Financial Reporting Advisory Group). (2024). *EFRAG's public consultation on two exposure drafts on sustainability reporting standards for SMEs*. Accessed March 18,

2024, from https://www.efrag.org/News/Public-479/EFRAGs-public-consultation-on-two-Exposure-Drafts-on-sustainability-r

Elkington, J. (1993). *Coming clean: The rise and rise of the corporate environmental report*, Business.

Elkington, J. (1998). *Cannibals with forks: The triple bottom line of sustainability*. New Society Publishers.

Elkington, J. (1999). *Triple bottom-line reporting: Looking for balance*, "Australian CPA", (March), pp. 19–21.

Epstein, M. J. (1996). *Measuring corporate environmental performance: Best practices for costing and managing an effective environmental strategy*. Irwin.

Erlingsson, C., & Brysiewicz, P. (2017). A hands-on guide to doing content analysis. *African Journal of Emergency Medicine, 7*(3), 93–99. https://doi.org/10.1016/j.afjem.2017.08.001

European Commission. (2022). *Targeted consultation on the functioning of the ESG ratings market in the European Union and on the Consideration of ESG Factors in Credit Ratings*. Summary Report, August 2022. Accessed March 11, 2024, from https://commission.europa.eu/system/files/2022-08/2022-esg-ratings-summary-of-responses_en.pdf

European Commission. (2001a). *Annual EU accounts and reports*. Accessed March 18, 2024, from https://commission.europa.eu/strategy-and-policy/eu-budget/how-it-works/annual-lifecycle/assessment/annual-accounts-and-reports_en

European Commission. (2001b). *Promoting a European framework for Corporate Social Responsibility*. Accessed March 07, 2024, from https://ec.europa.eu/commission/presscorner/api/files/document/print/en/doc_01_9/DOC_01_9_EN.pdf

European Commission. (2024). *Public consultation on new Sustainability Reporting Standards for SMEs under the CSRD*. Accessed March 20, 2024, from https://green-business.ec.europa.eu/news/public-consultation-new-sustainability-reporting-standards-smes-under-csrd-2024-02-29_en

European Federation of Pharmaceutical Industries and Associations. (2016). *SMEs in Europe: Biopharmaceutical SMEs and their role in the industry*. Accessed March 20, 2024, from https://www.efpia.eu/about-medicines/development-of-medicines/smes-in-europe/

European Union. (2014). *Directive 2014/95/EU of the European Parliament and of the Council of 22 October 2014 amending Directive 2013/34/EU as regards disclosure of non-financial and diversity information by certain large undertakings and groups*. Accessed March 07, 2024, from https://eur-lex.europa.eu/eli/dir/2014/95/oj

European Union. (2022). *Directive (EU) 2022/2464 of the European Parliament and of the Council of 14 December 2022 amending Regulation (EU) No 537/2014, Directive 2004/109/EC, Directive 2006/43/EC and Directive 2013/34/EU, as regards corporate sustainability reporting*. Accessed March 07, 2024, from https://eur-lex.europa.eu/legal-content/EN/TXT/?uri=CELEX:32022L2464

European Union. (2023). *Commission Delegated Regulation (EU) 2023/2772 of 31 July 2023 supplementing Directive 2013/34/EU of the European Parliament and of the Council as regards sustainability reporting standards*. Accessed March 07, 2024, from https://eur-lex.europa.eu/legal-content/en/TXT/?uri=CELEX:32023R2772

Eurostat. (2023). *2021: EU micro & small businesses employed 75.8 million*. Accessed March 20, 2024, from https://ec.europa.eu/eurostat/web/products-eurostat-news/w/ddn-20231109-2

FISMA (Directorate-General for Financial Stability, Financial Services and Capital Markets Union). (2023). *The Commission adopts the European Sustainability Reporting Standards*. Accessed March 07, 2024, from https://finance.ec.europa.eu/news/commission-adopts-european-sustainability-reporting-standards-2023-07-31_en

Flower, J. (2015). The international integrated reporting council: A story of failure. *Critical Perspectives on Accounting, 27*, 1–17. https://doi.org/10.1016/j.cpa.2014.07.002

Gholami, A., Murray, P. A., & Sands, J. (2022). Environmental, social, governance & financial performance disclosure for large firms: Is this different for SME firms? *Sustainability, 14*(10), 6019. https://doi.org/10.3390/su14106019

Gilchrist, D. J., West, A., & Zhang, Y. (2023). Barriers to the usefulness of non-profit financial statements: Perspectives from key internal stakeholders. *Australian Accounting Review, 33*, 188–202. https://doi.org/10.1111/auar.12401

Gjergji, R., Vena, L., Sciascia, S., & Cortesi, A. (2021). The effects of environmental, social and governance disclosure on the cost of capital in small and medium enterprises: The role of family business status. *Business Strategy and the Environment, 30*, 683–693. https://doi.org/10.1002/bse.2647

Global Reporting Initiative (GRI). (2016). *Universal Standard*. Accessed March 19, 2024, from https://www.globalreporting.org/standards/standards-development/universal-standards/

Global Reporting Initiative (GRI). (n.d.-a). *Collaborations for sustainability*. Accessed March 19, 2024, from https://www.globalreporting.org/public-policy/

Global Reporting Initiative (GRI). (n.d.-b). *Sector program*. Accessed March 19, 2024, from https://www.globalreporting.org/standards/sector-program/

Global Reporting Initiative (GRI). (n.d.-c). *A short introduction to the GRI standards*. Accessed March 19, 2024, from https://www.globalreporting.org/media/wtaf14tw/a-short-introduction-to-the-gri-standards.pdf

Gutiérrez-Hernández, P., & Abásolo-Alessón, I. (2021). The health care sector in the economies of the European Union: An overview using an input–output framework. *Cost Effectiveness and Resource Allocation, 19*(1), 4. https://doi.org/10.1186/s12962-021-00258-8

Henisz, W., Koller, T., & Nuttall, R. (2019). *Five ways that ESG creates value. McKinsey Quarterly*. Accessed March 20, 2024, from https://www.mckinsey.com/business-functions/strategy-and-corporatefinance/our-insights/five-ways-that-esg-creates-value#

Hillary, R. (2000). *Small and medium-sized enterprises and the environment business imperatives*. Accessed March 20, 2024, from https://nibmehub.com/opac-service/pdf/read/Small%20and%20medium-sized%20enterprises%20and%20the%20environment%20%20business%20imperatives%20by%20Hillary-%20Ruth.pdf

Hillary, R. (2004). Environmental management systems and the smaller enterprise. *Journal of Cleaner Production, 12*(6), 561–569. https://doi.org/10.1016/j.jclepro.2003.08.006

IFRS Foundation. (2024a). *Corporate reporting dialogue*. Accessed March 19, 2024, from https://integratedreporting.ifrs.org/corporate-reporting-dialogue/

IFRS Foundation. (2024b). *Integrated thinking*. Accessed March 19, 2024, from https://integratedreporting.ifrs.org/integrated-thinking/

Integrated Reporting. (2011). *Towards integrated reporting: Communicating value in the 21st century*. Accessed March 19, 2024, from https://integratedreporting.ifrs.org/wp-content/uploads/2011/09/IR-Discussion-Paper-2011_single.pdf

International Integrated Reporting Council. (2020). *10 Years of the IIRC*. Accessed March 19, 2024, from https://integratedreporting.ifrs.org/10-years/10-years-summary/

Johnson, M. P., & Schaltegger, S. (2016). Johnson and Schaltegger. *Journal of Small Business Management, 54*(3), 481–505. https://doi.org/10.1111/jsbm.12154

Karsam, K., Sasmita, J., Hudin, N. S., Rainanto, B. H., Solihin, S., & Noor, B. (2024). The influence of firm size, corporate governance, green accounting disclosure on sustainability report disclosure and financial performance and its impact on value company. *International Journal of Trends in Accounting Research, 3*(2), 84–93. https://doi.org/10.54951/ijtar.v3i2.401

Kim, S., & Li, Z. (2021). Understanding the impact of ESG practices in corporate finance. *Sustainability, 13*(7), 3746. https://doi.org/10.3390/su13073746

Kotsantonis, S., Pinney, C., & Serafeim, G. (2016). ESG integration in investment management: Myths and realities. *Journal of Applied Corporate Finance, 28*(2), 10–16.

Laine, M., Tregidga, H., & Unerman, J. (2021). Sustainability reporting, history, frameworks and regulation. In M. Laine, H. Tregidga, & J. Unerman (Eds.), *Sustainability accounting and accountability* (pp. 83–102). Routledge.

Lawrence, S. R., Collins, E., Pavlovich, K., & Arunachalam, M. (2006). Sustainability practices of SMEs: The case of NZ. *Business Strategy and the Environment, 15*(4), 242–257. https://doi.org/10.1002/bse.533

Mathews, M. R. (1995). Social and environmental accounting: A practical demonstration of ethical concern? *Journal of Business Ethics, 14*(8), 663–671. http://www.jstor.org/stable/25072684

MedTech Europe. (2021). *European health and SMEs – big challenges, small & medium sized solutions*. Accessed March 20, 2024, from https://www.medtecheurope.org/resource-library/european-health-and-smes-big-challenges-small-medium-sized-solutions/

Mio, C., Fasan, M., & Costantini, A. (2020). Materiality in integrated and sustainability reporting: A paradigm shift? *Business Strategy and the Environment, 29*, 306–320. https://doi.org/10.1002/bse.2390

Nguyen, P. T. T., & Kimura, A. (2023). Disclosure characteristics of annual reports and information asymmetry: Evidence from foreign firms listed on the US stock exchange. *Finance Research Letters, 54*, 103776. https://doi.org/10.1016/j.frl.2023.103776

O'Dwyer, B., Humphrey, C., & Rowbottom, N. (2024). From institutional integration to institutional demise: The disintegration of the International Integrated Reporting Council (IIRC). *Critical Perspectives on Accounting, 99*, 102699. https://doi.org/10.1016/j.cpa.2023.102699

ORSE (Observatoire de la Responsabilité Sociétale des Entreprises). (2007). *Guide to Sustainability Analysis Organisations—Profiles*. Report, December. Accessed March 11, 2024, from https://www.orse.org/fichier/2806

Ozkan, S., Romagnoli, S., & Rossi, P. (2023). A novel approach to rating SMEs' environmental performance: Bridging the ESG gap. *Ecological Indicators, 157*, 111151. https://doi.org/10.1016/j.ecolind.2023.111151

Panta, H., & Panta, A. (2023). Organizational capital and readability of financial reports. *Finance Research Letters, 55*(Part A), 103895. https://doi.org/10.1016/j.frl.2023.103895

Petrescu, A. G., Bîlcan, F. R., Petrescu, M., Holban Oncioiu, I., Türkeş, M. C., & Căpuşneanu, S. (2020). Assessing the benefits of the sustainability reporting practices in the top Romanian companies. *Sustainability, 12*, 3470. https://doi.org/10.3390/su12083470

Picas, S., Reis, P., Pinto, A., & Abrantes, J. L. (2021). Does tax, financial, and government incentives impact long-term Portuguese SMEs' sustainable company performance? *Sustainability, 13*(21), 11866. https://doi.org/10.3390/su132111866

Qian, A., & Zhu, D. (2019). Financial report similarity and the likelihood of administrative punishment: Based on the empirical evidence of textual analysis. *China Journal of Accounting Studies, 7*(2), 147–169. https://doi.org/10.1080/21697213.2019.1642604

Revell, A., & Blackburn, R. (2007). The business case for sustainability? An examination of small firms in the UK's construction and restaurant sectors. *Business Strategy and the Environment, 16*(6), 404–420. https://doi.org/10.1002/bse.499

Revell, A., Stokes, D., & Chen, H. (2010). Small businesses and the environment: Turning over a new leaf? *Business Strategy and the Environment, 19*(4), 273–288. https://doi.org/10.1002/bse.628

Rimmel, G. (2020). Accounting for sustainability – historical development of the field. In G. Rimmel (Ed.), *Accounting for sustainability* (pp. 3–15). Routledge.

Roncalli, T. (2024). *Handbook of sustainable finance*. Working Papers hal-04370824. Accessed March 10, 2024, from https://ideas.repec.org/p/hal/wpaper/hal-04370824.html

Rossi, P., & Candio, P. (2023). The independent and moderating role of choice of non-financial reporting format on forecast accuracy and ESG disclosure. *Journal of Environmental Management, 345*, 118891. https://doi.org/10.1016/j.jenvman.2023.118891

Santos, P. G.d., Albuquerque, F., Rodrigues, M. A. B., & Morais, A. I. (2022). The views of stakeholders on mandatory or voluntary use of a simplified standard on non-financial information for SMEs in the European Union. *Sustainability, 14*(5), 2816. https://doi.org/10.3390/su14052816

Saraite-Sariene, L., Alonso-Cañadas, J., Galán-Valdivieso, F., & Caba-Pérez, C. (2020). Non-financial information versus financial as a key to the stakeholder engagement: A higher education perspective. *Sustainability, 12*, 331. https://doi.org/10.3390/su12010331

Seidel, M., Seidel, R., Tedford, D., Cross, R., Wait, L., & Hämmerle, E. (2009). Overcoming barriers to implementing environmentally benign manufacturing practices: Strategic tools for SMEs. *Environmental Quality Management, 18*(1), 37–55. https://doi.org/10.1002/tqem.20214

Setyaningsih, S., Widjojo, R., & Kelle, P. (2023). Challenges and opportunities in sustainability reporting: A focus on small and medium enterprises (SMEs). *Cogent Business & Management*. Advance online publication. https://doi.org/10.1080/23311975.2023.2298215.

Shalhoob, H., & Hussainey, K. (2023). Environmental, Social and Governance (ESG) disclosure and the Small and Medium Enterprises (SMEs) sustainability performance. *Sustainability, 15*(1), 200. https://doi.org/10.3390/su15010200

Singh, K., & Pillai, D. (2022). Corporate governance in small and medium enterprises: A review. *Corporate Governance, 22*(1), 23–41. https://doi.org/10.1108/CG-10-2020-0470

Stubbs, W., & Higgins, C. (2018). Stakeholders' perspectives on the role of regulatory reform in integrated reporting. Journal of Business Ethics, 147, 489–508.

Van Wyk, M., & Els, G. (2023). The relevance of integrated reporting in future standard setting of the International Sustainability Standards Board. *Frontiers in Sustainability, 4*. https://doi.org/10.3389/frsus.2023.1218985

Venturelli, A., Caputo, F., Leopizzi, R., & Pizzi, S. (2019). The state of art of corporate social disclosure before the introduction of non-financial reporting directive: A cross country analysis. *Social Responsibility Journal, 15*(4), 409–423. https://doi.org/10.1108/SRJ-12-2017-0275

Wachira, M. M., Berndt, T., & Romero, C. M. (2020). The adoption of international sustainability and integrated reporting guidelines within a mandatory reporting framework: Lessons from South Africa. *Social Responsibility Journal, 16*(5), 613–629. https://doi.org/10.1108/SRJ-12-2018-0322

# Chapter 4
# Evolution of ESG and Corporate Performance in Health Care

## Background

Society is facing the consequences of climate change, depletion of natural resources, precarious working conditions, and corporate scandals. Consequently, the expectation regarding the environmental, social, and ethical responsibilities of companies has become increasingly relevant over time (Money & Schepers, 2007). Nowadays, various stakeholders are more aware and actively demand companies to adopt a more ethical approach (Dyduch & Krasodomska, 2017). Investors and stakeholders show significant interest in knowing where exactly the company invests its financial resources and how it conducts business activities, especially concerning environmental, social, and governance (ESG) activities. As a result, the demand for sustainability reporting and disclosure is growing, accompanied by an increasing demand for sustainable products and the search for information on associated risks, returns, and costs (Michelon & Parbonetti, 2012).

ESG activities were initially considered activities carried out only by a limited number of specialized investment funds. Today, sustainability is considered an investment philosophy adopted by a growing proportion of listed and non-listed companies. It is evident that investors now show a tendency to "align their investment decisions with their personal values," (Hartzmark & Sussman, 2019) making ESG considerations an integral part of common business process practices (Busch et al., 2016). Such a change is motivated by the desire to move from a short-term shareholder value creation policy to a more sustainable management strategy that takes into account the interests of diverse stakeholder groups.

Particularly for the health care industry, the Covid-19 pandemic has acted as a significant catalyst for sustainability. The crisis has highlighted the vulnerabilities of health care companies worldwide, pushing towards greater attention to social responsibility and the need to build a resilient and sustainable health care infrastructure. This has stimulated the adoption of ESG practices in the health care sector,



P. Candio, *Sustainability and Corporate Performance in Health Care*, CSR, Sustainability, Ethics & Governance,
https://doi.org/10.1007/978-3-031-63985-2_4

promoting positive change and creating opportunities for long-term innovation and growth and consequently corporate performance.

## Data Source

The sample of health care companies analyzed in Chaps. 4, 5 and 6 is part of the Euro STOXX 600 index, which is a stock index consisting of 600 prominent companies listed in European capital markets. This index includes large companies from 17 European countries, covering approximately 90% of the total market capitalization of the European stock market. The health care sector, represented by 55 listed companies, accounts for 9.2% of the total index composition.

To enable a quantitative assessment of the evolution of ESG and corporate performance, company data and information from the 2012–2022 period were extracted from the Refinitiv platform (Eikon, 2022) using the Datastream application for Microsoft Excel. Following retrieval from the platform, data were prepared for subsequent statistical analysis using STATA 15 software (StataCorp, 2019). An integral part of this evaluative process involved encoding the variables used for analysis, selected based on literature studies concerning the relationship between "sustainability," "non-financial reporting choices," and "corporate performance." The Appendix includes the variables selected for the development of the present and Chaps. 5 and 6, with an accurate description.

### *ESG Score Calculation*

Refinitiv's ESG scores represent the underlying structure of ESG data and enable the production of an assessment based on ESG performance data and actual company capabilities. This assessment integrates and considers sector materiality and biases related to company size. The methodology for assigning Refinitiv's ESG scores follows several key calculation principles (outlined below), which have been detailed in Chap. 3.

In brief, an overall ESG score is calculated, which is adjusted for relevant controversies faced by the company that may impact the proper conduct of business operations. Out of the 630 ESG measures calculated at the company level, a subset of 186 indicators comprising the most comparable and relevant measures for each sector fuels the evaluation process and enables the attainment of an overall company score (Fig. 4.1).

These 186 measures are grouped into 10 categories that define the scores related to the pillars and allow for the determination of the ESG score assigned at the end of the process. This assessment, therefore, clarifies the company's ESG performance, commitment, and effectiveness based on publicly reported information (see Table 4.1).

The category scores are grouped into the three ESG pillar scores: environmental, social, and governance. The individual scores for each ESG pillar are determined by a relative sum of category weights that vary for companies in the environmental and

| | | |
|---|---|---|
| ENVIRONMENTAL SCORE | Resource use | 20 |
| | Emissions | 28 |
| | Innovation | 20 |
| SOCIAL SCORE | Workforce | 30 |
| | Human rights | 8 |
| | Community | 14 |
| | Product responsibility | 10 |
| GOVERNANCE SCORE | Management | 35 |
| | Shareholders | 12 |
| | CSR strategy | 9 |
| | ***Total number of measures*** | ***186*** |

**Fig. 4.1** Refinitiv's ESG score measure composition. Fonte: Author's elaboration from page 6 https://www.lseg.com/content/dam/marketing/en_us/documents/methodology/refinitiv-esg-scores-methodology.pdf

**Table 4.1** ESG score categories

| ESG category | Code | Description |
|---|---|---|
| Resource use score | Resuse | Reflects a company's performance and its ability to reduce the use of materials, energy, or water, and to find more eco-efficient solutions by improving supply chain management. |
| Emissions reduction score | Emissions | Measures a company's commitment and effectiveness in reducing environmental emissions in its production and operational processes. |
| Innovation score | Envinn | Reflects a company's ability to reduce costs and environmental burdens for its customers, thereby creating new market opportunities through new environmental technologies and processes or ecologically designed products. |
| Workforce score | Workforce | Measures the effectiveness of a company in providing job satisfaction, a healthy and safe working environment, maintaining diversity and equal opportunities, as well as development opportunities for its workforce. |
| Human rights score | Humanrights | Measures a company's effectiveness in adhering to fundamental human rights conventions. |
| Community score | Commun | Measures a company's commitment to being a good corporate citizen, protecting public health, and adhering to corporate ethics. |
| Product responsibility score | Prodresp | Reflects a company's ability to produce quality goods and services, integrating customer health and safety, data integrity, and privacy. |
| Management score | Manag | Measures the commitment and effectiveness of a company in following corporate governance principles of best practices. |
| Shareholders score | Sharehold | Measures a company's effectiveness in fair treatment of shareholders and the use of anti-hostile devices. |
| CSR strategy score | CSRstrat | Reflects a company's practices to communicate and integrate economic (financial), social, and environmental dimensions into its daily decision-making processes. |

Source: https://www.lseg.com/content/dam/data-analytics/en_us/documents/methodology/lseg-esg-scores-methodology.pdf social categories. For the governance pillar, however, the weights remain constant. The pillar weights are normalized into percentages ranging from 0 to 100.

## Corporate Performance

Table 4.2 breaks down the industry by sector subgroups and shows how pharmaceutical and medical equipment production account for almost two-thirds of the number of listed companies involved in the health care business in Europe.

Based on the classification provided by the European Commission in terms of defining micro, small, and medium-sized enterprises based on the number of employees,[1] the sample under study only includes two small enterprises (with 10 to 49 employees) and 17 medium-sized enterprises (with 50 to 249 employees), with an average annual value range from 17 to 638,154. For the health care industry, a similar distribution is observed, with only one company classified as medium-sized (with $n = 234$ employees). The others, however, report exponentially higher values, up to $n = 263{,}819$ (Fig. 4.2).

A similar distribution is obtained when health care companies are ordered by the average total revenues value (Fig. 4.3).

On the other hand, Fig. 4.4 below shows how, on average, that listed health care companies have managed to generate proportionally higher profits compared to the rest of the companies listed in the Euro Stoxx 600 index and belonging to other sectors.

Additionally, this figure shows how during the Covid-19 pandemic period, particularly in 2021, these companies recorded exceptional profitability performance, reaching 42%, compared to the rest of the sample. In comparison and considering the return on total assets of the company (ROA), health care sector companies appear to have enjoyed significantly higher profitability (between 8% and 11%) compared to the rest of the sample (between 6% and 7%, Fig. 4.5).

**Table 4.2** Health Care industry sub sectors

| Industry Classification Benchmark (ICB) subsector name | Frequency | Percent | Cumulative |
|---|---|---|---|
| Biotechnology | 99 | 16% | 16% |
| Health Care Services | 22 | 4% | 20% |
| Medical Equipment | 165 | 27% | 47% |
| Medical Services | 11 | 2% | 49% |
| Medical Supplies | 99 | 16% | 65% |
| Pharmaceuticals | 209 | 35% | 100% |

[1] For further information regarding this recommendation by the Commission, the text of the document is available at the following link: https://eur-lex.europa.eu/legal-content/IT/TXT/?uri=celex%3A32003H0361

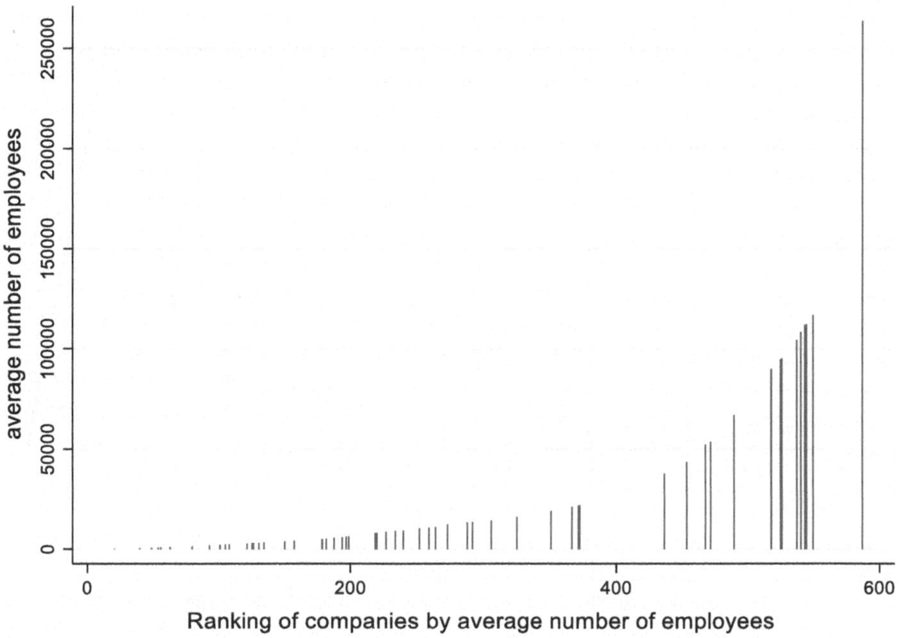


**Fig. 4.2** Ranking of health care companies by average number of employees

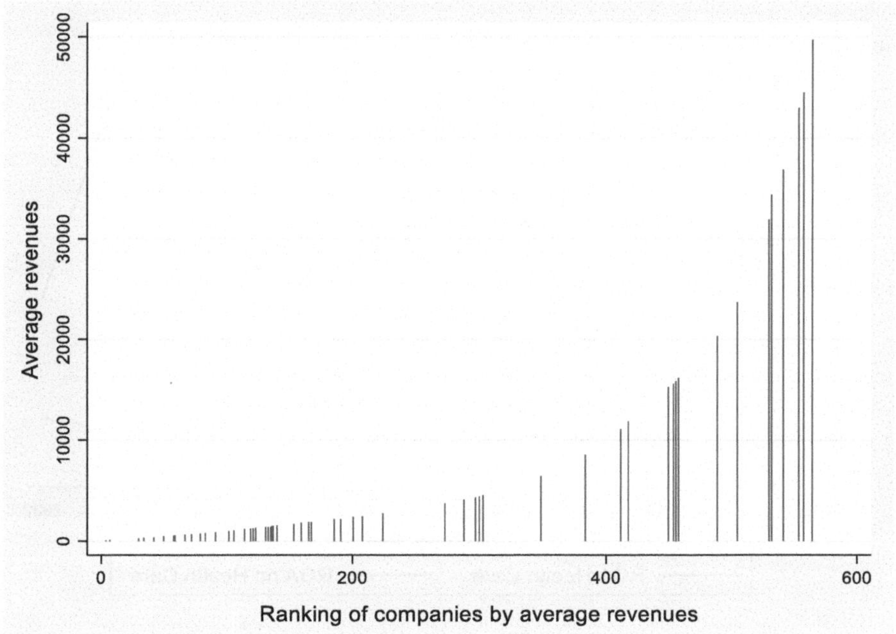


**Fig. 4.3** Ranking of health care companies by average revenues

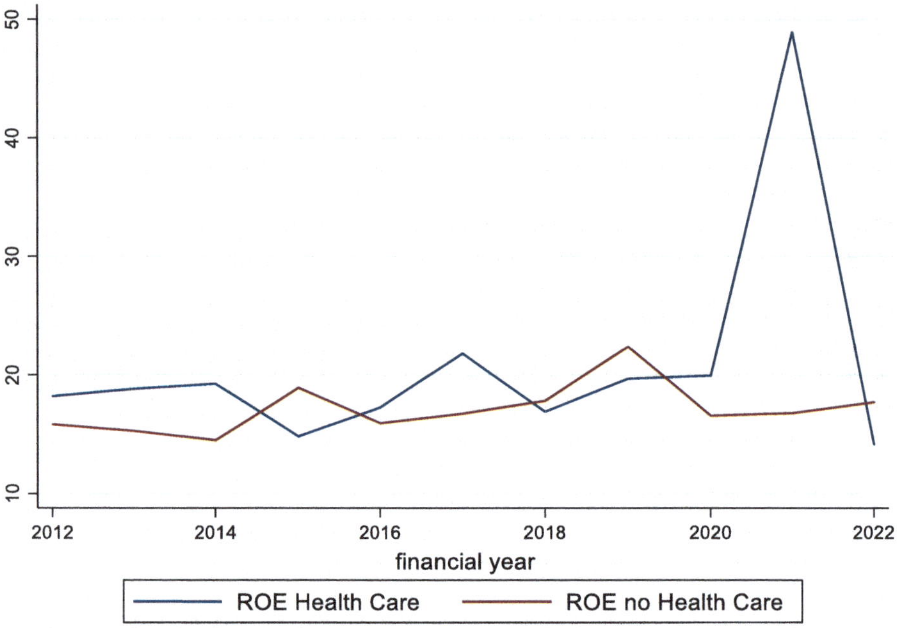


**Fig. 4.4** ROE Health Care and the rest of the sample

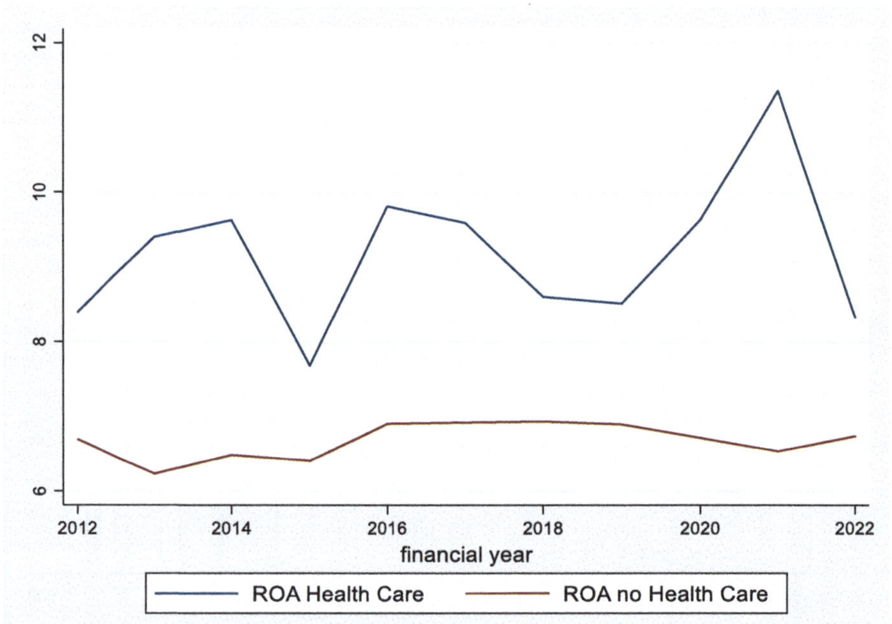


**Fig. 4.5** ROA Health Care and the rest of the sample

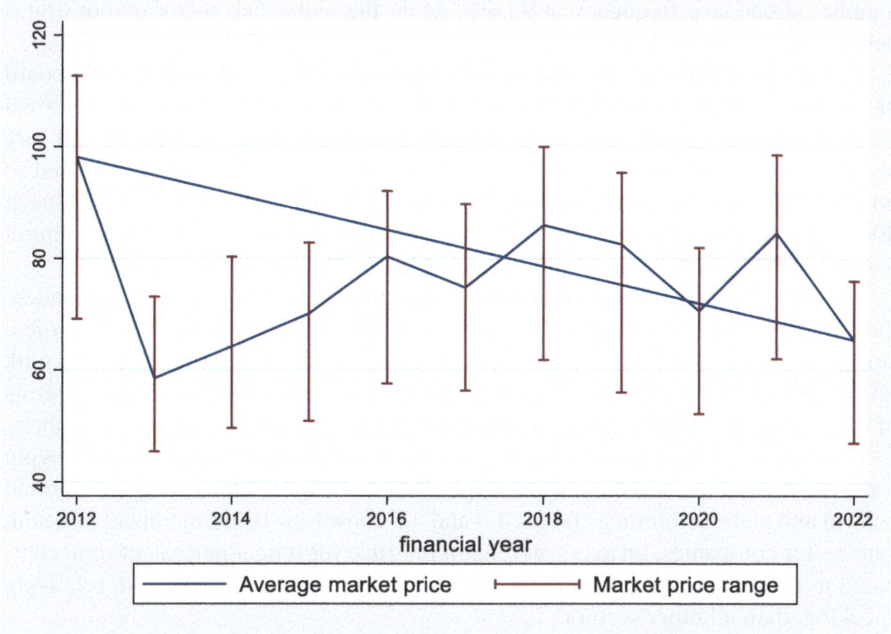


**Fig. 4.6** Trend in Health Care industry market price

Considering market value, which is the selling price based on the issuer company's future income prospects, Fig. 4.6 shows how for companies in the health care industry, the average stock price has experienced a decreasing trend over the last decade, dropping from an average price of EUR 80 to EUR 64 in 2022. It is worth noting the level of volatility associated with this aspect, with an average range of about 50% of the price, which, however, has gradually decreased over time to reach values of around 30%.

## ESG-Relevant Indicators

### *CSR and Board Characteristics*

The characteristics and activities carried out by a corporate board of directors can significantly influence the results stemming from the adoption of corporate social responsibility and, consequently, the implementation of sustainability in business processes. In this regard, among the eleven sectors considered within the Euro Stoxx 600 index, health care companies rank third from last with an annual average of 10.35 board members, between the real estate sector at 9.21 and companies dedicated to providing and managing citizen services at 12.75. Conversely, albeit with a relatively minor difference, health care companies rank last in terms of board member attendance frequency at 95.4%, while the real estate sector demonstrates being at the top with 96.9%.

Significant differences can also be noted regarding the composition of the board of directors, both in terms of the gender of the members and in terms of educational level. On average, about 30% of board members are female, with the health care sector at a minimum of 25% in 2022, and with marginal differences compared to other sectors. In contrast, the investigated sector presented a higher average value at 40%, ranging between 38% and 45% of board members with a different cultural background from where the company's registered office is located (Fig. 4.7).

Another important aspect of board composition is the presence of independent directors compared to executive members and their level of technical and professional readiness. As shown in Table 4.3, companies in the health care sector rank relatively high in a median position compared to companies in other sectors in terms of the proportion of independent board members as well as non-executive members.

If over 95% of the top 600 companies listed on European markets report having a board of directors with a consistent professional profile and having an internal control and audit committee, Tables 4.4 and 4.5 show how board members of health care sector companies, on average, remain in office for longer periods of time compared to most other sectors and how they are technically more prepared, relatively speaking, than all other sectors.

Regarding the actions conducted by companies in terms of sustainability reporting, Fig. 4.8 shows how sustainability activities carried out by companies, both

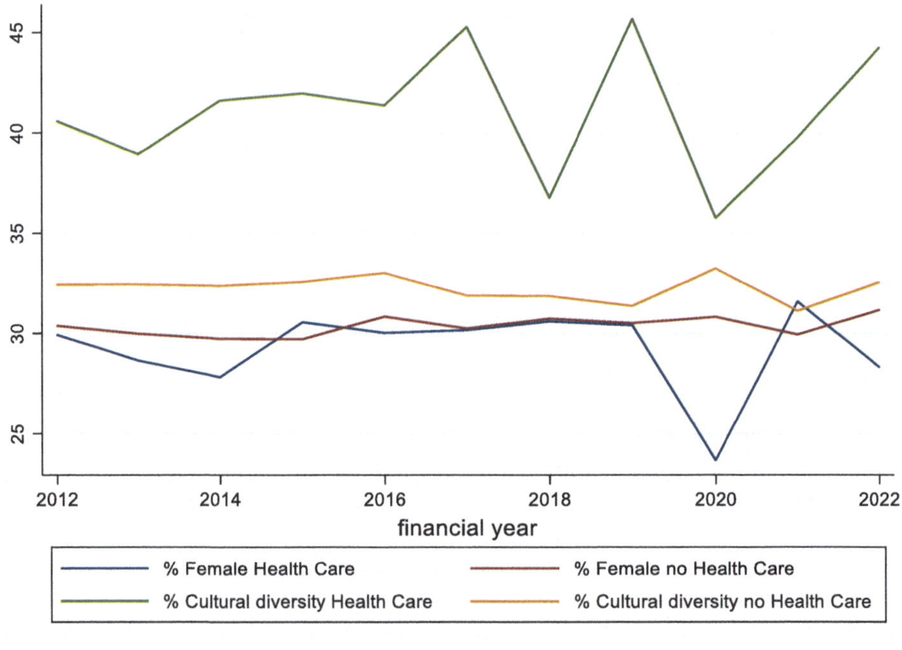


**Fig. 4.7** Comparative trends in board characteristics

**Table 4.3** Board level of independence

| Industry | % Independent board members | % Non-executive members |
|---|---|---|
| Basic Materials | 89.4% | 65.6% |
| Consumer Discretionary | 82.6% | 56.5% |
| Consumer Staples | 84.4% | 63.1% |
| Energy | 84.6% | 63.6% |
| Financials | 84.1% | 62.8% |
| Health Care | 86.3% | 65.1% |
| Industrials | 84.0% | 62.7% |
| Real Estate | 80.7% | 56.3% |
| Technology | 87.1% | 65.5% |
| Telecommunications | 90.2% | 66.0% |
| Utilities | 83.8% | 61.5% |

**Table 4.4** Board member average tenure

| Industry | % Average board member tenure (years) |
|---|---|
| Financials | 5.03 |
| Energy | 5.24 |
| Industrials | 5.42 |
| Utilities | 5.71 |
| Consumer Discretionary | 6.38 |
| Basic Materials | 6.61 |
| Technology | 6.62 |
| Health Care | 6.62 |
| Real Estate | 6.83 |
| Telecommunications | 6.83 |
| Consumer Staples | 7.08 |

**Table 4.5** Board technical background

| Industry | % Specific or strong financial background |
|---|---|
| Financials | 31.2% |
| Consumer Staples | 34.2% |
| Industrials | 35.4% |
| Real Estate | 36.0% |
| Energy | 37.0% |
| Telecommunications | 37.3% |
| Technology | 38.1% |
| Basic Materials | 38.2% |
| Utilities | 39.2% |
| Consumer Discretionary | 42.5% |
| Health Care | 43.1% |

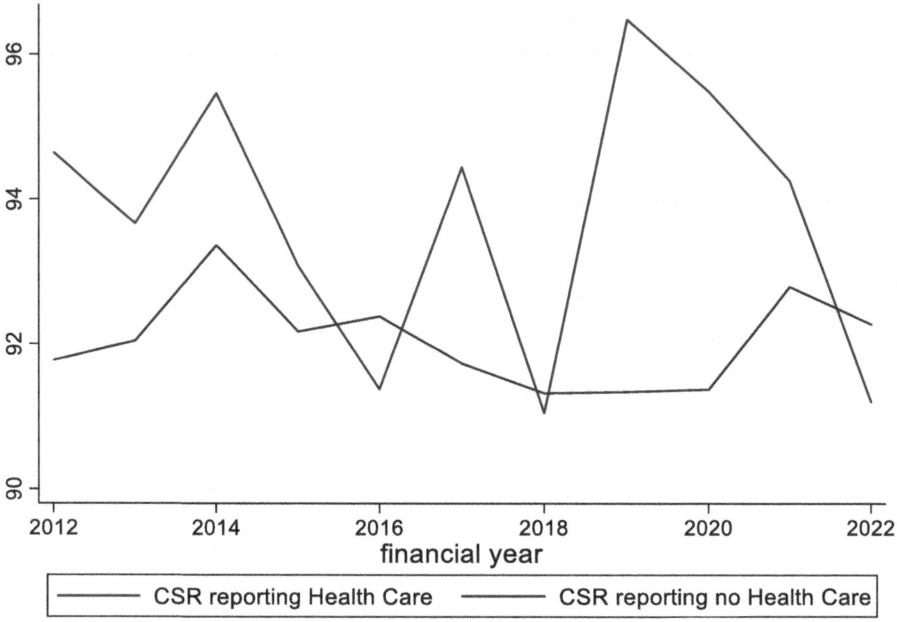


**Fig. 4.8** Trends in CSR reporting level

those belonging to the health care sector and those that do not, have been substantially reported within the reporting forms, ranging from a minimum of 90% to as high as 96.2%.

Regarding the choice made by listed companies to publish a sustainability report or dedicate the related reporting to a section in the annual financial statements, Fig. 4.9 shows how the use of the former reporting mode has increased from 40% in 2012 to over 70%, with non-significant differences between health care companies and those included in other sectors.

Similar proportions are evident in the choices regarding the integration of financial and non-financial factors within the management and analysis section included in the annual financial statements by listed companies in the health care sector and those belonging to other productive sectors (Fig. 4.10), with a median value of 25% for the first group compared to 30% for the second group of co.

Significant sector-level differences also emerge from the analysis of the choice of sustainability reporting framework used. In particular, health care companies show to be, on average, less inclined to adopt the Global Reporting Initiative framework,[2] with an average value of 40%, compared to 55% for the rest of the companies. Regarding corporate involvement and interaction with stakeholders,

[2] Further details regarding the Global Reporting Initiative (GRI) framework and dedicated reporting standards can be found on the website: https://www.globalreporting.org/how-to-use-the-gri--standards/get-started-with-reporting/

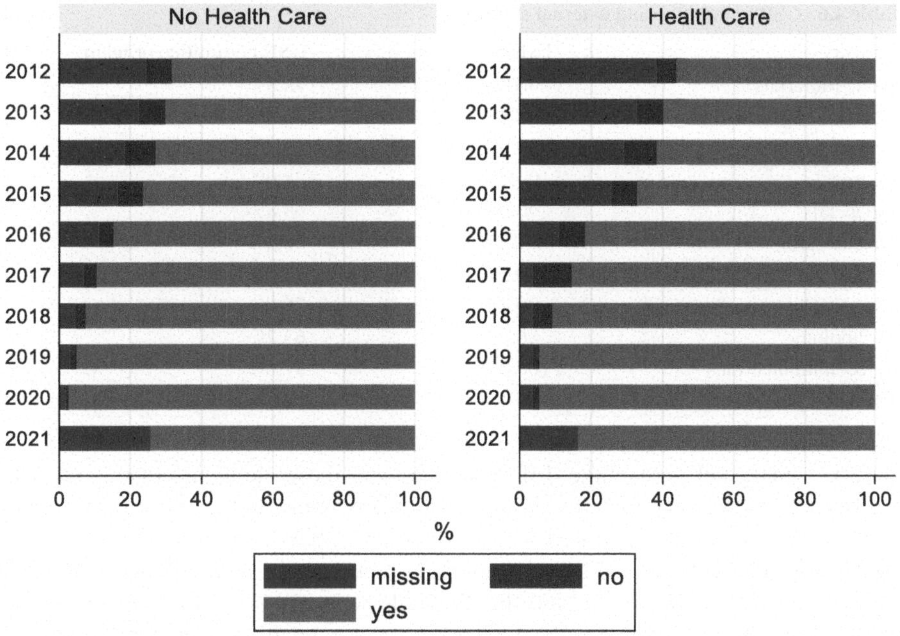


Fig. 4.9 CSR reporting framework

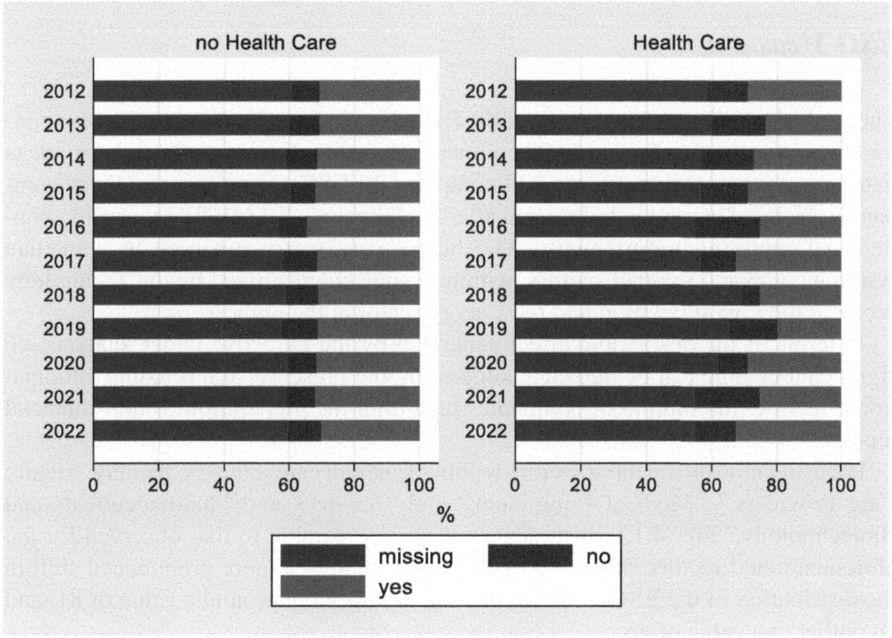


Fig. 4.10 Integration of financial and non-financial information

**Table 4.6** CSR committee and external auditor

| Industry | External auditor | CSR committee or team |
|---|---|---|
| Basic Materials | 90.8% | 78.8% |
| Consumer Discretionary | 94.8% | 80.4% |
| Consumer Staples | 90.9% | 89.4% |
| Energy | 94.1% | 78.0% |
| Financials | 90.6% | 68.8% |
| Health Care | 81.5% | 66.3% |
| Industrials | 90.3% | 78.6% |
| Real Estate | 90.7% | 76.0% |
| Technology | 86.9% | 63.2% |
| Telecommunications | 100.0% | 81.6% |
| Utilities | 100.0% | 80.8% |

there is a marginal percentage difference between these elements. In this regard, health care companies remain moderately behind compared to those in other sectors.

In contrast to what was found for stakeholders, Table 4.6 shows how the health care sector ranks last with 81.5% of companies reporting the presence of an external auditor for CSR activities, behind the Technology sector (86.9%). Conversely, at 66.3%, compared to 63.2% for the Technology sector, health care sector companies benefit more from the presence of a dedicated CSR committee or team.

## *ESG Measures*

The analysis of data available within the Refinitiv platform has allowed for examining the performance of the top 600 companies listed on European financial markets in terms of sustainability. Figure 4.11 compares the ESG performance of listed companies, grouped by sector, before and after the Directive 2014/95/EU regarding non-financial reporting requirements. The health care sector emerges in a median position compared to other sectors, within a range characterized by the Technology sector at the lowest position and the Energy sector at the highest.

In terms of the observable effect generated by the Directive under analysis, no significant change can be detected, suggesting the presence of a possible anticipation effect by this sample of companies regarding the preparation of non-financial reports.

Focusing now on the three sectors within the health care industry, namely "Health Care Providers," "Medical Equipment," and "Services and Pharmaceuticals and Biotechnology," Fig. 4.12 illustrates a pattern comparable to that observed for the aforementioned sectors, albeit the first subsector shows a more pronounced shift in the distribution of the ESG score, with a maximum score around a value of 81, and an outlier at a value of 45.

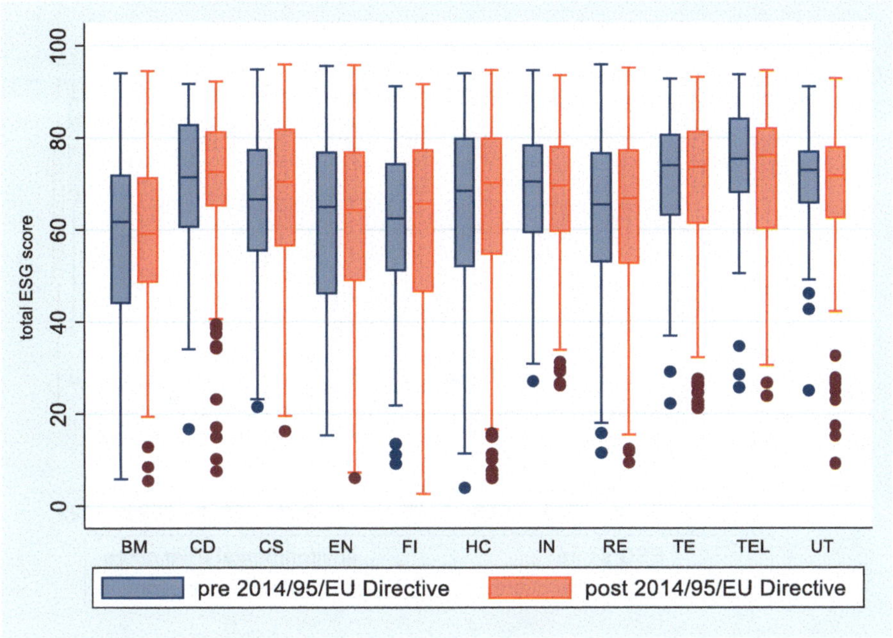


**Fig. 4.11** ESG score before and after 2014/95/EU Directive by industry

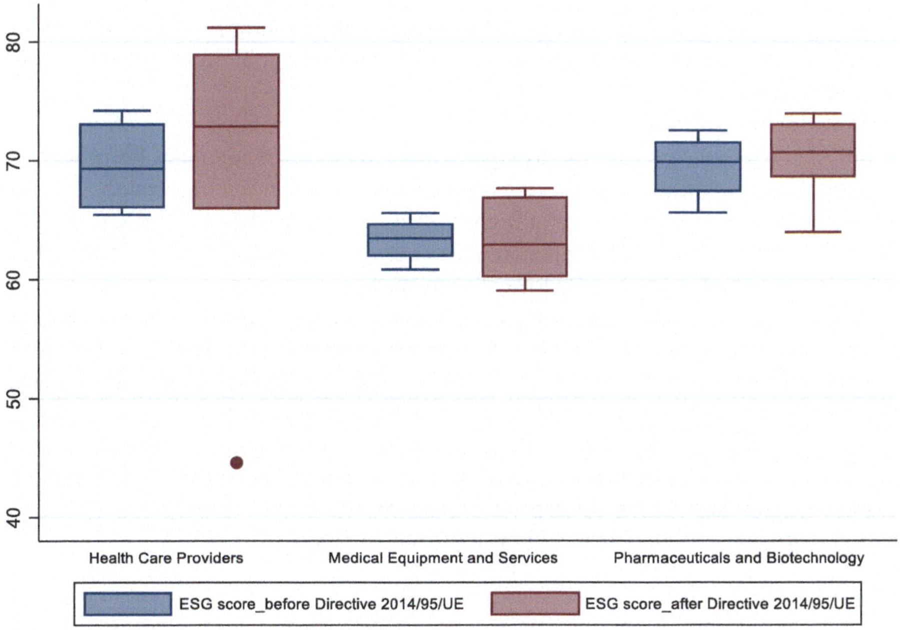


**Fig. 4.12** ESG score pre-post 2014/95/EU Directive by health care sector

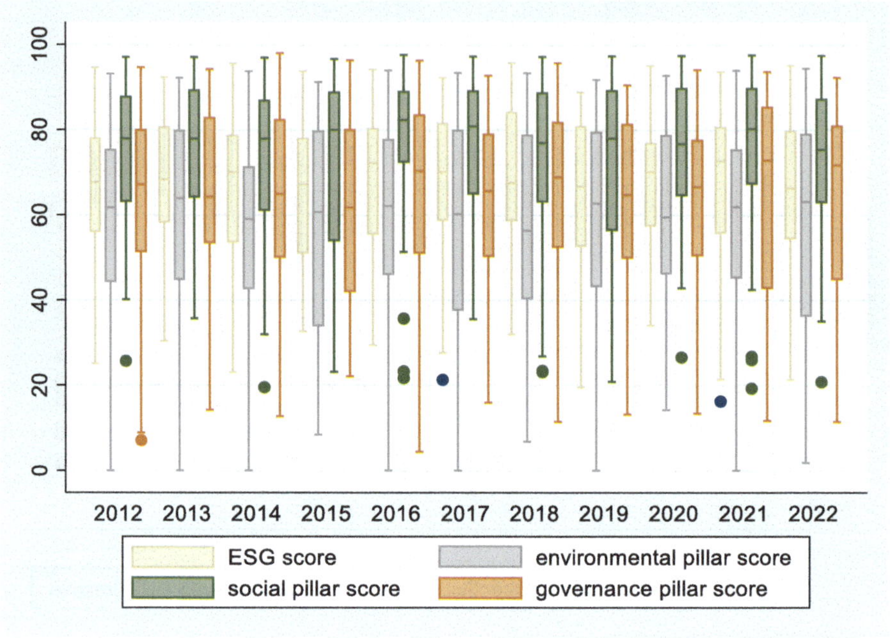


**Fig. 4.13** Temporal evolution of ESG score

In terms of temporal trends, Fig. 4.13 illustrates the evolution of the ESG performance of the health care sector categorized by standards in analysis and by assigned scores.

Both for the overall score and for the partial scores related to environmental, social, and governance pillars, no significant effects of change can be observed. Among the three pillars, the social aspect records a higher average value, with its minimum score values standing at higher positions compared to the other two pillars, while the environmental aspect remains comparatively lower, presenting a flatter distribution and reaching the lowest score values.

Finally, moving on to analyze in detail the constituent elements of the three ESG pillars and their temporal evolution concerning the implementation of Directive 2014/95/EU, in line with the previous illustrations, heterogeneity analyses indicate that there have been no significant changes in the average scoring of the ten categories constituting the three ESG pillars. Furthermore, companies belonging to the health care sector lag behind in terms of environmental innovation and defining a CSR strategy, while they excel in product responsibility and commitment, as well as in corporate capacity to adhere to best governance practices.

## Concluding Remarks

The top-ranking health care companies listed on European stock exchanges exhibit specific characteristics that are reflected in their CSR activities and, consequently, their ESG performance. The analysis presented in this chapter focusing on listed companies has shown that the health care industry is primarily composed of pharmaceutical companies, medical equipment manufacturers, and service providers. In Europe, this sector accounts for 9% of the total of listed companies, with only three companies with a headquarter in Italy. Furthermore, this sector is characterized by a particularly heterogeneous group of companies, both in terms of generated income and number of employees, with the majority of the sample being classed as large companies. Benchmarking analysis revealed that this sector has recorded significantly higher average profitability. It is noteworthy that the Covid-19 pandemic has indeed positively influenced the sector's profitability, contrary to the lower and marginal performance levels recorded for other industries.

Regarding corporate governance characteristics and CSR activities, the health care sector ranks among the lowest in terms of number of board members and their average attendance. In terms of board composition, companies in the health care sector show a distribution comparable to the rest of the sample in terms of proportion of female members, welcoming members with more diverse backgrounds, longer average tenure, and more technical expertise compared to those in other sectors. The analysis of the health care sector has also shown a propensity towards non-financial reporting choices in line with other sectors. This includes the choice of reporting framework, integration of financial and non-financial data within the management and analysis section included in the annual financial statements, and involvement of stakeholders in the company's decision-making process.

A significant difference was found in terms of type of sustainability reporting adopted, with a lower proportion of companies opting for the Global Reporting Initiative system. The comparative analysis of sustainability performance conducted in this chapter has shown that the entry into force of Directive 2014/95/EU[3] regarding non-financial reporting obligations has not significantly influenced the ESG performance of the sample of companies listed on European stock exchanges, with marginal differences emerged between the various sectors considered. This serves to demonstrate how the companies under analysis adapted in advance to what would later be required by the relevant regulations, given the relative importance of sustainability issues for this type of company. Analyses of heterogeneity and temporal trends have illustrated that both the overall ESG score and the scores related to the three pillars (environmental, social, and governance) have not undergone significant changes. Among the three pillars, the social score appears to obtain the highest value and a more favorable distribution in the health care sector compared to the

[3] For further information regarding this Directive of the European Union, please consult the relevant text available on the official website of the European Union: https://eur-lex.europa.eu/legal-content/IT/TXT/PDF/?uri=CELEX:32014L0095

other two pillars. Delving into the categories of ESG scores, the analysis has shown that for the health care sector, environmental innovation and the definition of CSR strategy lag behind other sectors, while the opposite is true for product responsibility and commitment, as well as corporate governance practices.

In order to interpret the results presented in this chapter, it is necessary to recognize the underlying peculiarities of the metrics used. While financial performance measured with ROA, ROE, and average share value is essentially taken from balance sheet information and included in official financial reports of companies and financial markets, this is not necessarily the case for CSR activities and characteristics and sustainability performance. As shown by recent empirical analysis, there are substantial differences in how different data providers and metrics operate. Opting to use the Refinitiv platform implicitly includes a certain level of bias in the values presented, which depend on the methodology used by the platforms (Berg et al., 2022). This includes both the relative weight of the numerous CSR and ESG indicators and sources of information, as well as the process of converting that information from non-traditional information channels used to calculate various scores and metrics. That said, it is certainly worth noting that the Refinitiv platform is one of the most reliable and widely used sources for corporate accountability value analysis. Different articles published in the literature focusing on companies in the health care sector serve as confirmation (Kalia & Aggarwal, 2023; El Khoury et al., 2023). In conclusion, the comprehensive description provided here regarding the aspects of this industry precedes the next two chapters, which will focus on exploring the effect and relationships between reporting framework choices, CSR characteristics, ESG profiles of health care companies and economic performance (Chap. 5) and information asymmetry (Chap. 6).

## References

Berg, F., Kölbel, J. F., & Rigobon, R. (2022). Aggregate confusion: The divergence of ESG ratings. *Review of Finance, 26*(6), 1315–1344. https://doi.org/10.1093/rof/rfac033

Busch, T., Bauer, R., & Orlitzky, M. (2016). Sustainable development and financial markets: Old paths and new avenues. *Business and Society, 55*(3), 303–329.

Dyduch, J., & Krasodomska, J. (2017). Determinants of corporate social responsibility disclosure: An empirical study of polish listed companies. *Sustainability, 9*, 1934. https://doi.org/10.3390/su9111934

Eikon. (2022). *Environmental, social and governance scores from Refinitiv*. Retrieved from https://www.refinitiv.com/content/dam/marketing/en_us/documents/methodology/refinitiv-esg-scores-methodology.pdf

El Khoury, R., Nasrallah, N., & Toumi, A. (2023). ESG and performance in public healthcare companies: The role of disclosure and director liability. *Competitiveness Review, 33*(1), 203–221. https://doi.org/10.1108/CR-12-2021-0174

Hartzmark, S. M., & Sussman, A. B. (2019). Do investors value sustainability? A natural experiment examining ranking and fund flows. *The Journal of Finance, 74*(6), 2789–2837.

Kalia, D., & Aggarwal, D. (2023). Examining impact of ESG score on financial performance of healthcare companies. *Journal of Global Responsibility, 14*(1), 155–176. https://doi.org/10.1108/JGR-05-2022-0045

Michelon, G., & Parbonetti, A. (2012). The effect of corporate governance on sustainability disclosure. *Journal of Management and Governance, 16*(3), 477–509.

Money, K., & Schepers, H. (2007). Are CSR and corporate governance converging? A view from boardroom directors and company secretaries in FTSE100 companies in the UK. *Journal of General Management, 33*(2), 1–11.

StataCorp. (2019). *Stata statistical software: Release 16*. StataCorp LLC [program].

# Chapter 5
# ESG and Economic Performance in Health Care

## Background

The current trend towards sustainability is reshaping the landscape of financial and accounting practices, as evidenced by a growing emphasis on sustainable finance objectives embraced by institutional investors and individuals, who prioritize investment in companies demonstrating robust ESG performance (Migliorelli, 2021). The expectations placed on companies regarding their ESG performance have consequently escalated over the past two decades, imposing greater pressures on companies to exhibit heightened levels of corporate social responsibility (Gao et al., 2022).

In this context, the health care industry has the potential to make significant contributions to the achievement of sustainability objectives (Chotchoungchatchai et al., 2020). Health care organizations committed to the Green Deal are mandated to pursue five core objectives tailored to their specific aims. These objectives entail enhancing the well-being of healthcare recipients and providers, deepening awareness of the environmental and climatic impacts associated with human activities, striving to attain carbon neutrality by 2050 through the adoption of energy-efficient practices in infrastructure, transportation, and procurement, with a particular focus on leveraging renewable energy sources, and ensuring responsible prescription and distribution of medications to minimize unnecessary consumption (European Commission, 2020). As a result, health care corporate managers are increasingly directing their attention towards ESG metrics, often initiating sustainability initiatives not only to bolster their company's reputation and economic performance, but also to serve their own vested interests (Broadstock et al., 2019).



P. Candio, *Sustainability and Corporate Performance in Health Care*, CSR, Sustainability, Ethics & Governance,
https://doi.org/10.1007/978-3-031-63985-2_5

## *Theoretical Frameworks*

Various theories have been employed in the academic literature to explain the relationship between CSR, and consequently ESG, and economic performance. The theoretical debate regarding the relationship between a company's economic and ESG performance has conventionally revolved much around two distinct theories: the agency theory and the stakeholder theory.

### Agency Theory

Agency theory, a cornerstone concept in corporate governance, provides valuable insights when applied to ESG practices. Within this framework, agency theory examines the dynamics between corporate managers, who hold decision-making authority, and shareholders who seek to maximize their returns. The theory posits that managers might prioritize personal gains over shareholder interests when engaging in ESG activities (Jensen & Meckling, 1976), potentially resulting in agency conflicts where managers pursue personal objectives rather than focusing on shareholder value creation. Indeed, a nuanced debate surrounds the rationale behind firms' adoption of sustainable practices. Some scholars contend that managers may utilize ESG activities to enhance their personal standing or attain non-financial benefits, potentially at the expense of shareholder value (Hillman et al., 2004). This perspective underscores the importance of aligning managerial incentives with shareholder interests to mitigate agency conflicts and ensure that ESG initiatives positively contribute to firm performance and sustainability outcomes. Conversely, advocates of ESG integration argue that responsible corporate conduct can yield long-term value for shareholders by mitigating risks, enhancing reputation, and fostering trust with stakeholders (Porter & Kramer, 2011).

In addition, engaging in ESG activities may necessitate firms to forgo projects that could potentially yield higher profits (Schuler & Cording, 2006). Allouche and Laroche (2005) assert that corporate social endeavors entail financial expenditures drawn from capital and other resources, placing the firm at a competitive disadvantage compared to less socially active counterparts. Thirdly, the managerial opportunism argument posits that managers allocate firm resources towards ESG activities to evade negative scrutiny and to compensate for or justify subpar financial performance, a practice commonly referred to as "window dressing": ESG initiatives are undertaken with the intention of gaining positive publicity to mask underlying weaknesses in performance (Gan & Ye, 2024). This viewpoint challenges the traditional agency theory narrative by proposing that ESG considerations can be in harmony with shareholder interests and foster sustainable value creation.

While the application of agency theory may highlight potential conflicts of interest between managers and shareholders in ESG decision-making (Donaldson & Davis, 1991), it also encourages exploration of how aligning incentives and promoting transparency can foster responsible corporate behavior. By embedding ESG considerations into strategic decision-making processes and governance frameworks, companies can effectively navigate agency challenges and harness the benefits of sustainable practices for long-term value creation. In practice, firms increasingly recognize the strategic importance of ESG integration not only for risk management but also for driving innovation, enhancing competitiveness, and meeting stakeholder expectations. Effective governance mechanisms that align managerial and value creation incentives are crucial to ensuring that ESG initiatives are integrated into core business strategies. By fostering a culture of accountability, transparency, and stakeholder engagement, organizations can bolster their resilience, reputation, and economic performance while making positive contributions to society and the environment. The convergence of agency theory and ESG presents a compelling opportunity for companies to address complex governance challenges and unlock sustainable value creation opportunities in a dynamic business environment.

## Stakeholder Theory

The stakeholder theory somewhat counters the agency theory by integrating a resource-based and a market-based view (Bhandari et al., 2022), linking the long-term economic survival of a business to its ability to balance and manage the interests of the company's multiple stakeholders (Freeman et al., 2010). This theory offers a comprehensive framework for understanding how corporate actions impact various stakeholders beyond just shareholders. Edward Freeman's stakeholder theory defines stakeholders as critical groups for a company's success, encompassing employees, suppliers, policymakers, environmental groups, society, media, and financial institutions, underscoring the importance of considering the interests of all stakeholders in decision-making processes. In examining the relationship between ESG and stakeholders' values, researchers have identified key factors pivotal in value creation. Integrated reporting, ESG practices, ESG ratings, sustainability reporting, and corporate governance practices are among the variables influencing stakeholder values. By aligning these factors with stakeholder interests, companies can bolster their competitive advantages, financial performance, and overall value creation for all stakeholders involved.

This inclusive approach aligns with the principles of stakeholder capitalism by offering a holistic perspective on how organizations can generate value for all stakeholders through ESG practices. By integrating stakeholder considerations into strategic decision-making processes and emphasizing transparency and accountability, companies can bolster their economic performance while contributing to ESG goals. This approach tends to be preferred in the research literature by virtue of the most recent theoretical developments that have identified a strategic and competitive dimension capable of generating value—and therefore higher performance—through better access to finance (Cochran, 2007), human resources management (Heal, 2005), risk and reputation management (Reinhardt et al., 2008), cost reduction through efficiency and sanctions avoidance (Greening & Turban, 2000).

## Review of the Empirical Literature

Considerable research has been undertaken to explore the influence of ESG scores on economic performance. Nevertheless, the existing body of evidence does not definitively establish whether activities related to corporate social responsibility and, consequently, ESG scores lead to enhanced performance for organizations, as studies have presented conflicting findings.

### *ESG and Economic Performance*

In the US context, Liu et al. (2020) utilized a multi-level regression approach and found an overall positive effect of environmental activities on Return on Assets (ROA). Conversely, Fahad and Busru (2021) identified evidence supporting a negative association between ESG and ROA when analyzing listed companies in India. Cheng and Wang (2011) examined the relationship between CSR activities and financial performance indicators in Chinese firms, revealing an overall positive impact. Han et al. (2016) used three distinct ESG scores as proxies for CSR activities in Korean listed companies and found heterogeneous results for individual ESG scores, with no indication that the social pillar influenced financial performance.

Recent studies in the European context have quantitatively explored this relationship. Agoraki et al. (2023) analyzed the association between a firm's ESG reputational risk and financial performance in European listed companies, finding that firms with lower reputational risk outperformed their counterparts. Hakimi et al. (2023) focused on European firms and discovered a positive and significant impact of ESG on financial performance, but only if the ESG score exceeded a certain threshold. Bifulco et al. (2023) analyzed European STOXX 600 index data and found a negative association between ESG score and market share price. Koundouri et al. (2022) investigated whether a relationship between strong ESG performance and favorable financial conditions could be established using a sample of top European companies, but found conflicting results among various financial indicators. Focusing on the food industry, Sandberg et al. (2023) found a modest positive association between ESG score and financial performance indicators using regression models applied to data from 83 companies over a four-year period.

These studies collectively highlight the complexity of the relationship between ESG activities and economic performance, with outcomes varying across regions, industries, and indicators.

#### Focusing on Health Care

To date, a relatively limited number of empirical studies have investigated the impact of ESG factors on economic performance, specifically focusing on the health care industry. El Khouri et al. (2022) explored how ESG scores predict accounting-based financial performance measures, including Return on Equity (ROE), Return on Assets (ROA), and Tobin's Q ratio, across 912 companies in 38 countries from 2012 to 2020. They found a concave impact of the environmental pillar on ROA, while no significant effects were observed for other ESG scores and financial performance metrics. In contrast, Agarwal et al. (2023) examined the association between ESG and market-based financial performance measures among 33 pharmaceutical companies in India from 2011 to 2020, revealing a significantly negative influence of total ESG score.

López-Toro et al. (2021) investigated the relationship between ESG indicators and financial performance among companies listed in the Nasdaq US Smart Pharmaceuticals Index ($n = 30$). They discovered a positive relationship between ESG indicators and financial ratios, suggesting that pharmaceutical companies' investments in sustainability and attention to controversies can enhance visibility and increase value. Kalia and Aggarwal (2023) analyzed the effect of total and individual ESG components on financial performance among 468 healthcare firms for the 2020 business year. Their study indicated that the relationship between ESG score and financial performance is context-dependent, with differing impacts in developed and developing economies. ESG activities positively impacted firm performance in developed economies but showed negative or insignificant relationships in developing economies.

Moreover, Gupta and Das (2022) conducted a meta-analytic review, and Galant and Cadez (2017) provided a comprehensive examination of measurement approaches, highlighting inconsistencies in study findings arising from variations in data settings, regression frameworks, autocorrelation effects, and choice of financial performance indicators. These factors underscore the importance of considering contextual nuances and methodological rigor when interpreting the relationship between ESG and financial performance in the healthcare sector.

## Research Methods

To empirically assess the proposition put forth in this chapter, which involves exploring the relationship between sustainability, articulated in terms of non-financial reporting, and the ESG score assigned by rating agencies and economic performance, the materials and data and analytical methods employed are described below.

### *Data Source and Sample*

The ESG, economic performance and all the other relevant company data used for the analysis presented in this chapter were sourced and downloaded from the Refinitiv platform (Eikon, 2022). Refinitiv's ESG scores are designed to objectively measure a company's ESG performance, commitment, and effectiveness based on data reported by the company. The three main ESG pillars (environmental, social, and governance) encompass ten ESG categories, namely, resource use, emissions reduction, innovation, workforce, human rights, community, product responsibility, management, shareholders, and CSR strategy score. Refinitiv also provides a comprehensive corrected ESG score that takes into account any significant ESG-related issues the company may have faced.

In order to enable a quantitative assessment of the relationships under study, various data and information pertaining to the 2012–2022 decade were extracted using the Datastream application for Microsoft Excel (Eikon, 2022). After downloading the data from the platform (specifically, 550 observations per company/year), they were prepared to enable subsequent statistical analysis. Variables were selected based on studies published in the literature concerning the relationship between sustainability and economic performance.

The sample of analyzed health care companies belonged to the Euro STOXX 600 index. This is a stock index comprised of 600 leading companies listed on capital markets within the European territory, consisting of a fixed number of 600 components including large companies with significant market capitalizations across 17 European countries. This composition covers approximately 90% of the total market capitalization of the European stock market. By concentrating on a singular albeit intricate geographical region, the research endeavors to mitigate the risk of the omitted-variable bias, a noteworthy challenge in investigations spanning multiple heterogeneous regions with diverse financial reporting regulations, jurisdictions, and customs. This methodological approach facilitates the simultaneous control of all fixed factors impacting both dependent and independent variables (De Jong et al., 2008).

The countries included in the index are Austria, Belgium, Denmark, Finland, France, Germany, Ireland, Italy, Luxembourg, Norway, the Netherlands, Poland, Portugal, the United Kingdom, Spain, Sweden, and Switzerland. The index is classified into 11 sectors, with the health care sector representing 9.2% of the total, with 55 listed companies. In this regard, the healthcare sector holds a nearly average position in terms of relative proportion to the number of companies comprising the index (Table 5.1).

**Table 5.1** Euro STOXX 600 index Industries

| Industry | *n* | % |
|---|---|---|
| Basic Materials | 40 | 6.7% |
| Consumer Discretionary | 86 | 14.3% |
| Consumer Staples | 46 | 7.7% |
| Energy | 22 | 3.7% |
| Financials | 105 | 17.5% |
| *Health Care* | 55 | 9.2% |
| Industrials | 136 | 22.7% |
| Real Estate | 31 | 5.2% |
| Technology | 32 | 5.3% |
| Telecommunications | 19 | 3.2% |
| Utilities | 28 | 4.7% |

## *Variables*

The dependent variable was economic performance, evaluated as net income per employee (netincempl), revenue per employee (revempl), and market value. Net income per employee is a metric that evaluates a company's productivity by dividing its net income by the number of employees. This ratio helps determine how effectively the business utilizes its workforce to generate profits. Revenue per employee is a metric of economic performance that evaluates the average profitability generated by each employee in a company. It is determined by dividing total revenue by its current employee count. This ratio helps organizations gauge their workforce's efficiency, productivity, and overall profitability. Market value was calculated by multiplying the number of outstanding shares and security price and it is a proxy measures used to assess the valuation and market perception of a company.

The primary set of explanatory variables were ESG scores: total, environmental, social, and governance pillars, as well as the ten score categories—which were treated as continuous variables. To account for determinants of economic performance, contextual and financial controls were included based on prior research (Lee & Suh, 2022). These controls encompassed various factors, namely: country of domicile; size, quantified by market value of equity, total assets, total revenues, market capitalization, and number of employees at the end of the financial year (all natural logarithms, ratio scale variable); leverage, calculated as the ratio of total debt to total assets at the end of the financial year; beta, indicating the volatility of the company's share compared to the broader market; book value per share, representing equity available to common shareholders relative to the number of shares outstanding; and market-to-book value, assessing the potential over or undervaluation of a company's share by comparing its market price with net assets.

## *Statistical Analysis*

Summary statistics were first employed to illustrate the characteristics of the company sample and explore variable relationships. Pearson's pairwise correlations were then calculated to assess the direction and strength of univariate linear associations. Linear regression specifications followed a forward stepwise approach, gradually building regression models. Models aimed to examine the relationship between ESG scores and economic performance, with βs representing the regression coefficient of interest, while controlling for relevant control factors (contextual and financial controls).

Specification 1 included only the primary explanatory variable, i.e. the total ESG score. Subsequent specifications incorporated the three ESG pillar scores (Model 2) and then all selected controls (Model 3). To compare predictor effects across multiple regression frameworks, all identified variables were retained in the final model specifications. Interaction terms were tested to assess potential effect modifications, and Wald tests were conducted to evaluate differences between categories. Statistical significance was set at a p-value <0.05 level. The analyses were performed using STATA 16 software (StataCorp, 2019).

## Relationship Between ESG and Economic Performance

Figures 5.1, 5.2, and 5.3 below depict the distribution of the top 55 health care companies in Europe listed in the Euro STOXX600 in terms of economic performance, as measured, respectively, by netincempl, revempl, and market value, and over time.

Concerning netincempl, an average of 34.8—significantly lower than that of the remaining pool of industries at 420.10—characterized healthcare companies, with a median value of 29.17 reaching 413.84 and 523.41% at the first and 99th percentiles of the distribution, respectively. In terms of temporal trends, the average performance values remained relatively unchanged, as shown in Fig. 5.1 which illustrates a relatively stable pattern of economic performance.

The revempl exhibits a rightly skewed distribution, with the majority of companies having values above 120 (Fig. 5.2). Over time, the average value has remained

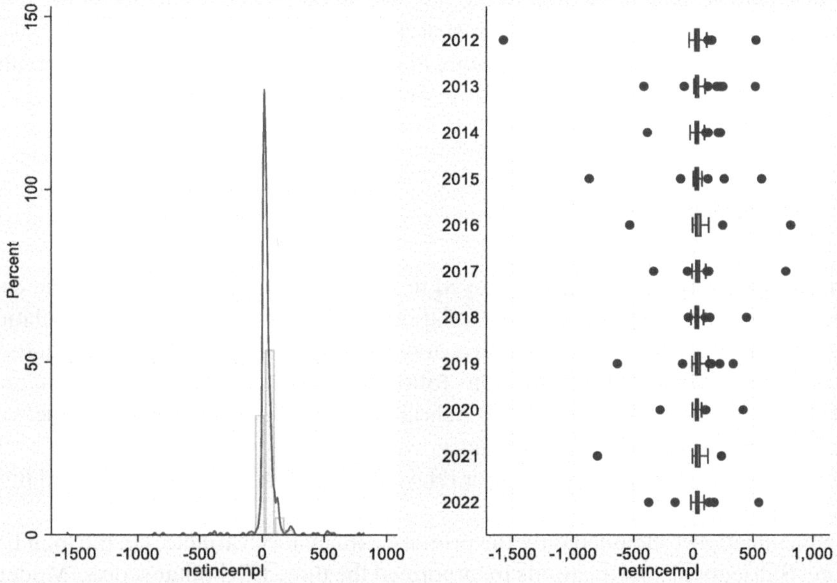


**Fig. 5.1** Distribution of health care companies by netincempl

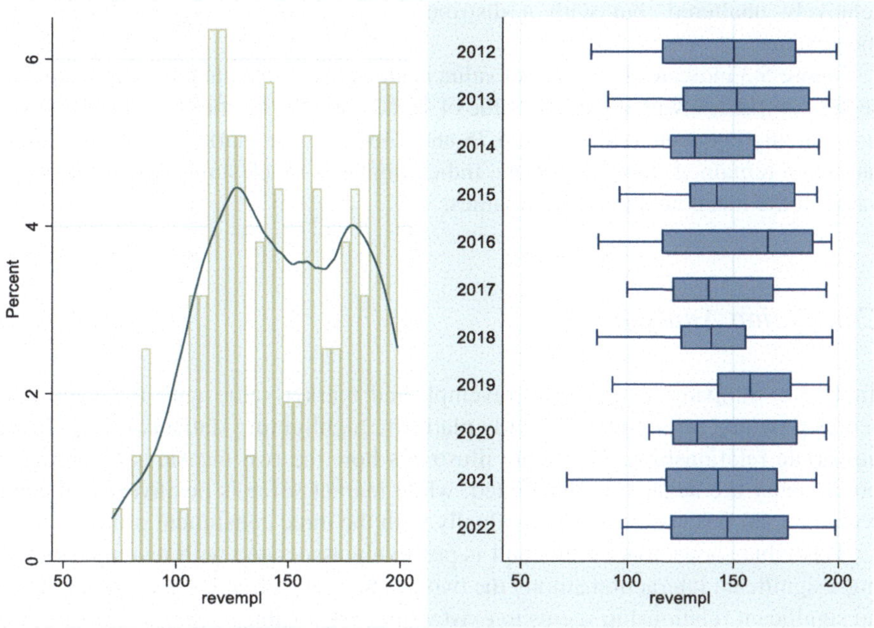


**Fig. 5.2** Distribution of health care companies by revempl

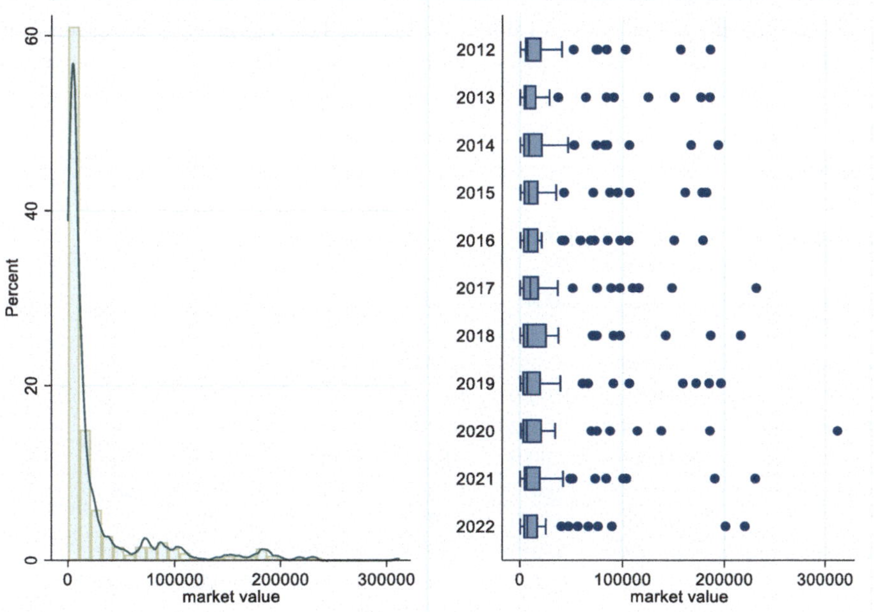


**Fig. 5.3** Distribution of health care companies by market value

relatively unaltered, but with a distribution gradually converging towards the mean value.

Figure 5.3 indicates how market value is unevenly distributed among health care sector companies, with a median value of 16,621 and the bottom 5% of the distribution ranging between values of 506.78 and 1360.57. The market value distribution has again remained stable over time, indicating an as much stable perception of the value of the listed healthcare companies.

## *Correlation Analysis*

Table 5.2 compares netincempl, revempl, and market value with the main ESG scores provided through the Refinitiv platform, highlighting statistically significant univariate relationships. This table illustrates how the two measures of employee productivity are negatively associated, while market value is positively associated with the total ESG score and individually with the three constituent pillars.

This table shows that netincempl is positively associated with revempl, indicating a significant interaction among the two productivity variables analyzed, whereas no significant relationship seems to exist with market value.

Shifting focus to the environmental pillar, Table 5.3 displays values perfectly consistent in terms of the direction of the effect (negative) showed in Table 5.2 concerning each of the three environmental score constituent categories, albeit with a relatively more modest effect.

Regarding the social score (Table 5.4), greater heterogeneity is observed compared to the environmental pillar, with all four categories being negatively

**Table 5.2** Pearson's correlations: ESG and economic performance

| Variables | (1) | (2) | (3) | (4) | (5) | (6) | (7) |
|---|---|---|---|---|---|---|---|
| (1) netincempl | 1.000 | | | | | | |
| (2) netincrev | 0.327* | 1.000 | | | | | |
| | (0.000) | | | | | | |
| (3) market value | −0.011 | 0.004 | 1.000 | | | | |
| | (0.410) | (0.739) | | | | | |
| (4) esgscore | −0.096* | −0.082* | 0.358* | 1.000 | | | |
| | (0.000) | (0.000) | (0.000) | | | | |
| (5) envscore | −0.094* | −0.062* | 0.318* | 0.809* | 1.000 | | |
| | (0.000) | (0.000) | (0.000) | (0.000) | | | |
| (6) socscore | −0.072* | −0.071* | 0.333* | 0.888* | 0.677* | 1.000 | |
| | (0.000) | (0.000) | (0.000) | (0.000) | (0.000) | | |
| (7) govscore | −0.055* | −0.053* | 0.223* | 0.701* | 0.341* | 0.420* | 1.000 |
| | (0.000) | (0.000) | (0.000) | (0.000) | (0.000) | (0.000) | |

***$p < 0.01$, **$p < 0.05$, *$p < 0.1$

**Table 5.3** Pearson's correlations: environmental and economic performance

| Variables | (1) | (2) | (3) | (4) | (5) | (6) |
|---|---|---|---|---|---|---|
| (1) netincempl | 1.000 | | | | | |
| (2) revempl | 0.310* | 1.000 | | | | |
| | (0.000) | | | | | |
| (3) market value | −0.011 | −0.007 | 1.000 | | | |
| | (0.410) | (0.578) | | | | |
| (4) resuse | −0.111* | −0.071* | 0.293* | 1.000 | | |
| | (0.000) | (0.000) | (0.000) | | | |
| (5) emiss | −0.068* | −0.024 | 0.295* | 0.727* | 1.000 | |
| | (0.000) | (0.082) | (0.000) | (0.000) | | |
| (6) envinn | −0.057* | −0.038* | 0.192* | 0.437* | 0.386* | 1.000 |
| | (0.000) | (0.007) | (0.000) | (0.000) | (0.000) | |

***$p < 0.01$, **$p < 0.05$, *$p < 0.1$

**Table 5.4** Pearson's correlations: social and economic performance

| Variables | (1) | (2) | (3) | (4) | (5) | (6) | (7) |
|---|---|---|---|---|---|---|---|
| (1) netincempl | 1.000 | | | | | | |
| (2) revempl | 0.310* | 1.000 | | | | | |
| | (0.000) | | | | | | |
| (3) market value | −0.011 | −0.007 | 1.000 | | | | |
| | (0.410) | (0.578) | | | | | |
| (4) workforce | −0.075* | −0.028* | 0.265* | 1.000 | | | |
| | (0.000) | (0.043) | (0.000) | | | | |
| (5) humrights | −0.046* | 0.004 | 0.254* | 0.439* | 1.000 | | |
| | (0.001) | (0.758) | (0.000) | (0.000) | | | |
| (6) commun | −0.041* | −0.017 | 0.283* | 0.448* | 0.501* | 1.000 | |
| | (0.003) | (0.229) | (0.000) | (0.000) | (0.000) | | |
| (7) prodresp | −0.079* | −0.031* | 0.216* | 0.447* | 0.379* | 0.372* | 1.000 |
| | (0.000) | (0.029) | (0.000) | (0.000) | (0.000) | (0.000) | |

***$p < 0.01$, **$p < 0.05$, *$p < 0.1$

associated with netincempl and positively with market value, whereas only two (workforce and prodresp) being negatively with revempl.

Finally, the scores of the ESG categories relating to governance (Table 5.5) exhibit association patterns broadly comparable to those found for the social score categories. Specifically, all three categories are negatively associated with netincempl, while csrstrat category score is positively and negatively associated with market value and revempl, respectively.

**Table 5.5** Pearson's correlations: governance and economic performance

| Variables | (1) | (2) | (3) | (4) | (5) | (6) |
|---|---|---|---|---|---|---|
| (1) netincempl | 1.000 | | | | | |
| (2) revempl | 0.310* | 1.000 | | | | |
| | (0.000) | | | | | |
| (3) market value | −0.011 | −0.007 | 1.000 | | | |
| | (0.410) | (0.578) | | | | |
| (4) manag | −0.044* | −0.024 | 0.212* | 1.000 | | |
| | (0.002) | (0.089) | (0.000) | | | |
| (5) sharehold | −0.032* | −0.031* | 0.010 | 0.197* | 1.000 | |
| | (0.023) | (0.027) | (0.474) | (0.000) | | |
| (6) csrstrat | −0.057* | −0.058* | 0.243* | 0.371* | 0.118* | 1.000 |
| | (0.000) | (0.000) | (0.000) | (0.000) | (0.000) | |

***$p < 0.01$, **$p < 0.05$, *$p < 0.1$

**Table 5.6** Univariate linear models

| Variables | netinc/employee | Revenues/employee | Market value |
|---|---|---|---|
| Total ESG score | −40.421*** | −652.932*** | 540.111*** |
| | (5.833) | (176.471) | (19.626) |
| Environmental pillar score | −29.410*** | −583.397*** | 358.743*** |
| | (4.334) | (130.946) | (14.888) |
| Social pillar score | −26.319*** | −244.031 | 436.284*** |
| | (5.083) | (153.598) | (17.188) |
| Governance pillar score | −19.526*** | −406.267*** | 283.555*** |
| | (4.909) | (148.066) | (17.299) |
| ESG controversies score | 5.461 | 154.835 | −358.005*** |
| | (3.764) | (113.419) | (12.603) |
| Observations | 5150 | 5149 | 5156 |
| R-squared | <0.01 | <0.01 | 0.105 |

Standard errors in parentheses ***$p < 0.01$, **$p < 0.05$, *$p < 0.1$

## *Regression Analysis*

Table 5.6 presents the values of univariate linear regression between the main ESG scores and the three economic performance variables for companies in the healthcare sector. This analysis indicates a marked heterogeneity of association among the investigated dimensions.

Net income per employee is negatively influenced by the overall ESG score (mean −40.421, SD 5.833) and to a lesser degree by the three pillar scores, while it does not seem to hold any statistical relationship with the ESG controversies score which in fact exhibits a notably wide and negative effect (mean −358.005, SD 12.603), with market value, suggesting that an increase in controversies is associated with a lower perceived company value. No evidence is shown for this score to affect revempl, as it is not the social score.

The total ESG, environmental pillar, and governance pillar scores significantly predict negative and positive revenue per employee and market value, respectively.

In order to correct the univariate effects observed in Table 5.6, a multivariate analysis was conducted considering the influence that other financial factors of health care sector companies may have on economic performance. Tables 5.7 and 5.8 focus on netincempl and consider, respectively, the total ESG score and the three score pillars. The progressive adjustment—evident from the transition from Model 1 to Model 3—shows how the effect of the total ESG score was in part masked by the control variables considered, particularly leverage which measures the debt level of the company, indicating that an debt to equity ratio is independently associated with lower economic performance as measured by netincempl.

Table 5.8 illustrates how the social and governance pillars fail to predict netincempl, while the environmental score affects it negatively.

Focusing now on revempl, Table 5.9 reveals that the progressive adjustment for control factors does not show any significant effect of the total ESG score. However, this analysis also unveils the influence of volatility, compared to the other control variables considered. The respective heterogeneity analyses reveal no systematic differences amongst the three ESG pillars, where none of the pillars appear to independently predict revempl.

Table 5.10 shows how, in contrast to the estimation for netincempl, none of the pillar scores have an effect on economic performance as measured by revempl, as illustrated by the progressive model adjustments.

Finally, Tables 5.11 and 5.12 focus on market value. Market value appears to be positively influenced by the total ESG score as well as by the three pillar scores

**Table 5.7** Multivariable regression models: netincempl and total ESG score

| Variables | Model 1 | Model 2 | Model 3 |
|---|---|---|---|
| Total ESG score | −40.421*** | −39.895*** | −34.186*** |
| | (5.833) | (5.854) | (5.730) |
| Subsector | | −0.000 | −0.000 |
| | | (0.000) | (0.000) |
| Book value per share | | | 0.075 |
| | | | (0.260) |
| Leverage | | | −13.385** |
| | | | (6.342) |
| Beta | | | 306.009 |
| | | | (221.735) |
| Market price | | | −0.175 |
| | | | (0.529) |
| Constant | 3048.693*** | 3322.574*** | 2799.645*** |
| | (394.785) | (470.023) | (487.045) |
| Observations | 5150 | 5150 | 5034 |
| R-squared | 0.009 | 0.009 | 0.009 |

Standard errors in parentheses ***$p < 0.01$, **$p < 0.05$, *$p < 0.1$

**Table 5.8** Multivariable regression models: netincempl and ESG score pillars

| Variables | Model 1 | Model 2 | Model 3 |
|---|---|---|---|
| Environmental pillar score | −25.244*** | −24.540*** | −16.204*** |
| | (5.867) | (5.936) | (5.727) |
| Social pillar score | −2.639 | −2.963 | −7.752 |
| | (7.101) | (7.114) | (6.821) |
| Governance pillar score | −9.081* | −9.174* | −8.336 |
| | (5.378) | (5.380) | (5.200) |
| Subsector | | −0.000 | −0.000 |
| | | (0.000) | (0.000) |
| Book value per share | | | 0.072 |
| | | | (0.261) |
| Leverage | | | −13.597** |
| | | | (6.346) |
| Beta | | | 331.414 |
| | | | (222.754) |
| Market price | | | −0.158 |
| | | | (0.529) |
| Constant | 2770.386*** | 2978.497*** | 2594.535*** |
| | (405.884) | (486.278) | (501.488) |
| Observations | 5150 | 5150 | 5034 |
| R-squared | 0.010 | 0.010 | 0.009 |

Standard errors in parentheses ***$p < 0.01$, **$p < 0.05$, *$p < 0.1$

**Table 5.9** Multivariable regression models: revempl and total ESG score

| Variables | Model 1 | Model 2 | Model 3 |
|---|---|---|---|
| Total ESG score | −652.932*** | −642.027*** | 4.414 |
| | (176.471) | (177.101) | (56.410) |
| Subsector | | −0.000 | 0.000 |
| | | (0.000) | (0.000) |
| Book value per share | | | 0.537 |
| | | | (2.562) |
| Leverage | | | −78.847 |
| | | | (62.377) |
| Beta | | | 4808.312** |
| | | | (2180.626) |
| Market price | | | −1.473 |
| | | | (5.203) |
| Constant | 51,403.556*** | 57,071.008*** | −4397.631 |
| | (11,944.094) | (14,215.255) | (4792.733) |
| Observations | 5149 | 5149 | 5033 |
| R-squared | 0.003 | 0.003 | 0.002 |

Standard errors in parentheses ***$p < 0.01$, **$p < 0.05$, *$p < 0.1$

**Table 5.10** Multivariable regression models: revempl and ESG score pillars

| Variables | Model 1 | Model 2 | Model 3 |
|---|---|---|---|
| Environmental pillar score | −778.376*** | −769.511*** | 9.868 |
| | (177.097) | (179.210) | (56.311) |
| Social pillar score | 499.573** | 495.494** | 14.902 |
| | (214.391) | (214.778) | (67.069) |
| Governance pillar score | −313.024* | −314.198* | −29.362 |
| | (162.382) | (162.436) | (51.132) |
| Subsector | | −0.000 | 0.000 |
| | | (0.000) | (0.000) |
| Book value per share | | | 0.487 |
| | | | (2.563) |
| Leverage | | | −79.644 |
| | | | (62.405) |
| Beta | | | 4849.885** |
| | | | (2190.228) |
| Market price | | | −1.526 |
| | | | (5.205) |
| Constant | 43,954.226*** | 46,575.066*** | −3850.128 |
| | (12,266.155) | (14,690.200) | (4933.992) |
| Observations | 5149 | 5149 | 5033 |
| R-squared | 0.005 | 0.005 | 0.002 |

Standard errors in parentheses ***$p < 0.01$, **$p < 0.05$, *$p < 0.1$

**Table 5.11** Multivariable regression models: market value and total ESG score

| Variables | Model 1 | Model 2 | Model 3 |
|---|---|---|---|
| Total ESG score | 540.111*** | 552.573*** | 559.092*** |
| | (19.626) | (19.599) | (19.919) |
| Subsector | | −0.000*** | −0.000*** |
| | | (0.000) | (0.000) |
| Book value per share | | | −6.898*** |
| | | | (0.906) |
| Leverage | | | −40.192* |
| | | | (22.353) |
| Beta | | | −1549.055** |
| | | | (779.292) |
| Market price | | | 16.217*** |
| | | | (1.823) |
| Constant | −18,267.081*** | −11,961.858*** | −11,069.678*** |
| | (1328.120) | (1575.133) | (1695.271) |
| Observations | 5156 | 5156 | 5073 |
| R-squared | 0.128 | 0.137 | 0.150 |

Standard errors in parentheses ***$p < 0.01$, **$p < 0.05$, *$p < 0.1$

**Table 5.12** Multivariable regression models: market value and ESG score pillars

| Variables | Model 1 | Model 2 | Model 3 |
|---|---|---|---|
| Environmental pillar score | 182.888*** | 206.107*** | 208.138*** |
| | (19.903) | (20.020) | (20.096) |
| Social pillar score | 242.947*** | 232.692*** | 234.897*** |
| | (23.954) | (23.857) | (23.830) |
| Governance pillar score | 116.593*** | 113.735*** | 117.938*** |
| | (18.212) | (18.113) | (18.215) |
| Subsector | | −0.000*** | −0.000*** |
| | | (0.000) | (0.000) |
| Book value per share | | | −6.891*** |
| | | | (0.903) |
| Leverage | | | −42.846* |
| | | | (22.279) |
| Beta | | | −1928.758** |
| | | | (779.768) |
| Market price | | | 15.954*** |
| | | | (1.817) |
| Constant | −18,545.837*** | −11,657.826*** | −10,475.796*** |
| | (1359.709) | (1620.961) | (1736.884) |
| Observations | 5156 | 5156 | 5073 |
| R-squared | 0.134 | 0.144 | 0.157 |

Standard errors in parentheses ***$p < 0.01$, **$p < 0.05$, *$p < 0.1$

where the social dimensions ranks first in terms of magnitude, closely followed by the environmental score and the governance score slightly lagging behind.

However, unlike for the two productivity measures, the control variables selected play a significant adjustment role and emerge as independent predictors of economic performance as measured by market value, particularly beta and leverage which negatively and positively affect performance, respectively, though with the former playing a more central role in terms of effect magnitude.

## Concluding Remarks

The current study has examined the influence of sustainability scores, gauged through ESG performance and its constituent environmental, social, and corporate governance pillars, on the economic performance of health care companies listed on European stock exchanges. Overall, the findings suggest that ESG score can effectively predict economic performance when measured in terms of net income per employment and market value, although with opposite effect directions. Upon delving into the underlying heterogeneity of the overall score, the analyses have unveiled some disparities in effect among the three ESG pillars where the environmental pillar appeared to play a more prominent role in predicting economic performance, relative to the other two pillars.

A recent study conducted by Kalia and Aggarwal (2023) has addressed a similar topic within the health care sector, drawing from a sample of 468 companies spanning a multinational context encompassing 33 countries in the year 2020. The authors suggest that the relationship between ESG scores and corporate performance cannot be generalized. While engaging in ESG activities yields positive impacts on the performance of healthcare companies in developed economies, this relationship may prove negative or insignificant in developing economies. This study contributes to the existing body of evidence by providing insights tailored to the European context over the past decade, thereby bolstering the internal validity of the findings, which warrant consideration within comparable legislative and regulatory frameworks.

The implications of this study are potentially significant for policymakers and practitioners seeking to enhance ESG activities and non-financial reporting disclosures. Moreover, a nuanced understanding of the impact of each pillar within the ESG score holds considerable utility for management and strategic decision-making within the healthcare sector. The significant effects ascertained of ESG activities on economic performance are poised to motivate boards of directors and management to escalate sustainability initiatives.

However, it is important to acknowledge some limitations. This study was based on a pooled-based regression approach and does not assert a causal relationship between ESG and economic performance. In fact, a reverse relationship could exist (Chams et al., 2021), as financially stronger companies may possess greater resources to invest in sustainability activities and disseminate related disclosures to financial markets. Additionally, these analyses are based on a single dataset, which, while being widely used, assigns scores to pillars and subcategories differently and derives pertinent information through varied processes and modes compared to other company data providers (Berg et al., 2022). This could significantly impact the results, necessitating future studies to validate these findings. Furthermore, this study contributes to the evidence base through empirical work drawn from a sample of publicly traded healthcare companies, predominantly of large scale, thereby limiting the generalizability of results to smaller and medium-sized enterprises, which face relatively higher sustainability reporting costs and are subject to different regulatory environments. Future studies will prove beneficial in this regard, particularly following the recent CSRD regulations, which will mandate a sample of these companies, if listed, to comply with sustainability reporting obligations, thereby making analyzable data available in the near future.

## References

Agarwal, B., Gautam, R. S., Jain, P., Rastogi, S., Bhimavarapu, V. M., & Singh, S. (2023). Impact of environmental, social, and governance activities on the financial performance of Indian health care sector firms: Using competition as a moderator. Journal of Risk and Financial Management, 16(2),109. https://doi.org/10.3390/jrfm16020109

Agoraki, M.-E. K., Giaka, M., Konstantios, D., & Patsika, V. (2023). Firms' sustainability, financial performance, and regulatory dynamics: Evidence from European firms. *Journal of International Money and Finance, 131*, 102785. ISSN 0261-5606. https://doi.org/10.1016/j.jimonfin.2022.102785
Allouche, J., & Laroche, P. (2005). A meta-analytical investigation of the relationship between corporate social and financial performance. *Revue de Gestion des Ressources Humaines, 57*(1), 18–41.
Berg, F., Koelbel, J. F., & Rigobon, R. (2022). Aggregate confusion: The divergence of ESG ratings. *Review of Finance, 26*(6), 1315–1344.
Bhandari, K. R., Ranta, M., & Salo, J. (2022). The resource-based view, stakeholder capitalism, ESG, and sustainable competitive advantage: The firm's embeddedness into ecology, society, and governance. *Business Strategy and the Environment, 31*, 1525–1537. https://doi.org/10.1002/bse.2967
Bifulco, G. M., Savio, R., Paolone, F., & Tiscini, R. (2023). The CSR Committee as moderator for the ESG score and market value. *Corporate Social Responsibility and Environmental Management, 30*(6), 3231–3241. https://doi.org/10.1002/csr.2549
Broadstock, D. C., Managi, S., Matousek, R., & Tzeremes, N. G. (2019). Does doing 'good' always translate into doing 'well'? An eco-efficiency perspective. Business Strategy and the Environment, 28(6), 1199–1217. https://doi.org/10.1002/bse.2311
Chams, N., García-Blandón, J., & Hassan, K. (2021). Role Reversal! Financial performance as an antecedent of ESG: The moderating effect of total quality management. *Sustainability, 13*, 7026. https://doi.org/10.3390/su13137026
Chen, H., & Wang, X. (2011). Corporate social responsibility and corporate financial performance in China: An empirical research from Chinese firms. *Corporate Governance, 11*(4), 361–370. https://doi.org/10.1108/14720701111159217
Chotchoungchatchai, S., Marshall, A. I., Witthayapipopsakul, W., Panichkriangkrai, W., Patcharanarumol, W., & Tangcharoensathien, V. (2020). Primary health care and sustainable development goals. *Bulletin of the World Health Organization, 98*(11), 792–800. https://doi.org/10.2471/BLT.19.245613
Cochran, P. L. (2007). The evolution of corporate social responsibility. *Business Horizons, 50*(6), 449–454.
De Jong, A., Kabir, R., & Nguyen, T. T. (2008). Capital structure around the world: The roles of firm- and country-specific determinants. *Journal of Banking & Finance, 32*(9), 1315–1344.
Donaldson, L., & Davis, J. H. (1991). Stewardship theory or agency theory: CEO governance and shareholder returns.
Eikon. (2022). *Environmental, social and governance scores from Refinitiv*. Retrieved from https://www.refinitiv.com/content/dam/marketing/en_us/documents/methodology/refinitiv-esg-scores-methodology.pdf
El Khoury, R., Nasrallah, N., Harb, E., & Hussainey, K. (2022, June 20). Exploring the performance of responsible companies in G20 during the COVID-19 outbreak. *Journal of Cleaner Production, 354*, 131693. https://doi.org/10.1016/j.jclepro.2022.131693
European Commission (2020). Shaping Europe's digital future. [Available from https://eur-lex.europa.eu/legalcontent/en/TXT/?uri=CELEX%3A52020DC0067, accessed 06.03.2024].
Fahad, P., & Busru, S. A. (2021). CSR disclosure and firm performance: Evidence from an emerging market. *Corporate Governance: The International Journal of Business in Society, 21*(4), 553–568. https://doi.org/10.1108/CG-05-2020-0201
Freeman, R. E., Harrison, J. S., Wicks, A. C., Parmar, B. L., & De Colle, S. (2010). *Stakeholder theory: The state of the art.*
Galant, A., & Cadez, S. (2017). Corporate social responsibility and financial performance relationship: A review of measurement approaches. *Economic Research-Ekonomska Istraživanja, 30*(1), 676–693. https://doi.org/10.1080/1331677X.2017.1313122
Gan, K., & Ye, S. (2024). Window dressing in impression management: Does negative media coverage drive corporate green production? *Sustainability, 16*(2), 861. https://doi.org/10.3390/su16020861

Gao, W., Li, M., & Zou, C. (2022). *Analysis of the impact of ESG on corporate financial performance under the epidemic based on static and dynamic panel data*. Wireless Communications and Mobile Computing, 2022.

Greening, D. W., & Turban, D. B. (2000). Corporate social performance as a competitive advantage in attracting a quality workforce. *Business & Society, 39*(3), 254–280.

Gupta, J., & Das, N. (2022). Multidimensional corporate social responsibility disclosure and financial performance: A meta-analytical review. *Corporate Social Responsibility and Environmental Management, 29*(4), 731–748. https://doi.org/10.1002/csr.2237

Hakimi, A., Boussaada, R. and Karmani, M. (2023), "Corporate social responsibility and firm performance: a threshold analysis of European firms", European Journal of Management and Business Economics, Vol. ahead-ofprint No. ahead-of-print. https://doi.org/10.1108/EJMBE-07-2022-0224

Han, J. J., Kim, H. J., & Yu, J. (2016). Empirical study on the relationship between corporate social responsibility and financial performance in Korea. *AJSSR, 1*, 61–76. https://doi.org/10.1186/s41180-016-0002-3

Heal, G. (2005). Corporate social responsibility: An economic and financial framework. *The Geneva Papers on Risk and Insurance-Issues and Practice, 30*(3), 387–409.

Hillman, A. J., Keim, G. D., & Schuler, D. (2004). Corporate political activity: A review and research agenda. *Journal of Management, 30*(6), 837–857.

Jensen, M. C., & Meckling, W. H. (1976). Theory of the firm: Managerial behavior, agency costs and ownership structure. *Journal of Financial Economics, 3*(4), 305–360.

Kalia, D., & Aggarwal, D. (2023). Examining impact of ESG score on financial performance of healthcare companies. *Journal of Global Responsibility, 14*(1), 155–176.

Koundouri, P., Pittis, N., & Plataniotis, A. (2022). The impact of ESG performance on the financial performance of European area companies: An empirical examination. *Environmental Sciences Proceedings, 15*(1), 13. https://doi.org/10.3390/environsciproc2022015013

Lee, M. T., & Suh, I. (2022). Understanding the effects of Environment, Social, and Governance conduct on financial performance: Arguments for a process and integrated modelling approach. *Sustainable Technology and Entrepreneurship, 1*(1), 100004.

Liu, F., Meng, L., Zhao, Y., et al. (2020). The influence of the corporate social responsibility disclosures on consumer brand attitudes under the impact of COVID-19. *Frontiers of Business Research in China, 14*, 28. https://doi.org/10.1186/s11782-020-00096-0

López-Toro, A. A., Sánchez-Teba, E. M., Benítez-Márquez, M. D., & Rodríguez-Fernández, M. (2021). Influence of ESGC indicators on financial performance of listed pharmaceutical companies. *International Journal of Environmental Research and Public Health, 18*(9), 4556. https://doi.org/10.3390/ijerph18094556

Migliorelli, M. (2021). What do we mean by sustainable finance? Assessing existing frameworks and policy risks. *Sustainability, 13*(2), 975. https://doi.org/10.3390/su13020975

Porter, M. E., & Kramer, M. R. (2011). *Creating shared value*. Harvard Business Review. Accessed February 23, 2024, from https://www.communitylivingbc.ca/wp-content/uploads/2018/05/Creating-Shared-Value.pdf

Reinhardt, F. L., Stavins, R. N., & Vietor, R. H. (2008). *Corporate social responsibility through an economic lens*. Accessed February 23, 2024, from https://scholar.harvard.edu/files/stavins/files/reinhardt_stavins_vietor_reep.pdf

Sandberg, H., Alnoor, A., & Tiberius, V. (2023). Environmental, social, and governance ratings and financial performance: Evidence from the European food industry. *Business Strategy and the Environment, 32*(4), 2471–2489. https://doi.org/10.1002/bse.3259

Schuler, D. A., & Cording, M. (2006). A corporate social performance-corporate financial performance behavioral model for consumers. *Academy of Management Review, 31*(3), 540–558. https://doi.org/10.5465/amr.2006.21318916

StataCorp. (2019). *Stata Statistical Software: Release 16*. StataCorp LLC [program].

# Chapter 6
# ESG and Information Asymmetry in Health Care

## Background

Information asymmetry occurs when one party involved in a financial transaction or decision-making process possesses a superior level and or quality of information compared to the other party (Rossi, 2019). In the corporate context, this often occurs between the management and external stakeholders, such as investors, creditors, or analysts. This phenomenon can manifest in various ways: (1) one party may have access to financial information that is not fully disclosed to the other party, resulting in a partial understanding of the financial status and performance of the company; (2) one party might receive financial information earlier than the other, thereby gaining a temporary advantage in making informed decisions; (3) discrepancies in the precision and reliability of information available to different parties; (4) management may choose to selectively disclose information, emphasizing positive aspects and downplaying negatives, thereby creating a distorted perception.

Information asymmetry among firms is generally considered a critical determinant of negative outcomes in financial performance. When investors do not receive reliable information to assess a company's operations, they may demand a higher interest rate, consequently leading to a higher cost of capital provision (Lambert et al., 2012). Moreover, the increasing complexity in evaluating a company in the absence or limitation of information could distort its realistic market valuation, increasing the risk of hostile takeovers (Grossman & Hart, 1981).

Over the past two decades, growing interest from large companies in ESG impacts has prompted a shift from traditional financial reporting to alternative modalities, particularly sustainability reporting and integrated reporting (Flower, 2015). While sustainability reports cater to a wide range of stakeholders and are published separately from purely financial accounting information, integrated reports offer an alternative approach by integrating financial values, operational data, and sustainability information to create short-, medium-, and long-term value



P. Candio, *Sustainability and Corporate Performance in Health Care*, CSR, Sustainability, Ethics & Governance,
https://doi.org/10.1007/978-3-031-63985-2_6

(Stubbs & Higgins, 2018). Such a significant change in reporting practices has sparked debate on the relative usefulness of different reporting forms and frameworks for stakeholders, particularly in terms of information asymmetry, especially following the recent implementation of the European Directive on non-financial reporting, which mandates large European companies to produce corporate reports with information on policies, risks, and performance regarding social responsibility, environment, human rights, and anti-corruption.[1] In such a context, the empirical study conducted in this chapter explores the mitigating effect of ESG information disclosed directly by companies to engage and align investors (Garcia-Sanchez & Noguera-Gamez, 2017).

## Theoretical Frameworks

From a theoretical standpoint, the disclosure of non-financial information and reports should reduce information asymmetry and consequently impact investors' judgments and behaviors by provision of higher-quality and more relevant information. This, in turn, would positively affect financial analysts' ability to forecast future returns with greater precision (Verrecchia, 2001). The theoretical debate regarding the relationship between information asymmetry and ESG performance has frequently revolved around two key theories: signaling theory and value relevance theory.

### *Signaling Theory*

Signaling theory examines the dynamics between two parties in the presence of information asymmetry (Spence, 2002). Prior to George A. Akerlof's introduction of the concept of information asymmetry in the academic discourse, economic models operated under the assumption of perfect information (Stiglitz, 2001). However, Akerlof (1970) challenged this assumption. He showed that information asymmetry existed in markets using the example of the used car market. In this market, sellers have better information regarding the quality of the vehicle than buyers, which results in suboptimal decisions and consequently market outcomes. Building on this work, Michael Spence (1973) further investigated the so-called market signaling effect by showing how job applicants can reduce information asymmetry with employers by signaling their profile quality through investments in education.

[1] For further information, please consult the relevant text available on the website: https://eur-lex.europa.eu/legal-content/EN/TXT/PDF/?uri=CELEX:32014L0095.

Since these seminal studies, the literature on signaling theory has been widely applied across various disciplines, including business economics, corporate governance, and human resource management (Connelly et al., 2011). However, to date, the application of signaling theory in the context of sustainability remains relatively limited, particularly regarding its focus on signaling mechanism. As part of their recommendations for future research, Connelly et al. (2011) proposed studying ESG initiatives, where these should signal a firm's credibility and expectation that it will improve its sustainability-related performance.

The currently limited research on ESG disclosure in the context of sustainability can be leveraged as it could provide evidence towards the signaling role of ESG disclosure and simultaneously introduce targets as a viable medium for signaling information of intent. The primary elements of signaling are the signaler, the signal, and the receiver (Connelly et al., 2011). First, the firm is the signaler, which is the party holding information unavailable to the receiver(s). Second, ESG disclosure is the signal, which is the information shared with the receiver(s). Third, investors are the receiver, which is the party lacking but desiring information. In management literature, particularly in entrepreneurship and strategy research, investors are often used as the receiving party.

The sustainability-related information signaled to the stakeholders can be divided into quality, focusing on characteristics, and information of intent, focusing on behavior (Stiglitz, 2001). As an example of the former, in Spence's (1973) study, job applicants indicated their ability through investing in their education, thereby signaling information of quality by showing the applicant's commitment and ability to put time and effort into completing that level of education. An example of signaling information of intent instead is when entrepreneurs approach angel investors to invest in their start-up, which is a type of behavior that could signal to investors that the entrepreneur intends to "cash out" (Elitzur & Gavious, 2003). Importantly, quality is both a characteristic of the signaler and the signal in this theoretical context, and a distinction can be made between high-quality entities that are able to truthfully signal information and low-quality entities that are not (Kirmani & Rao, 2000).

Taking the net-zero emission transition as a case in point, in an ideal world, all companies should eventually possess the ability to make such transition and thus no low-quality firms would exist. However, given past research (Gershon, 2013; Schumpeter, 1942; Rogers, 1995), industries and firms will have different levels of transition preparedness (Schoenmaker & Schramade, 2023). Hence, with respect to decarbonization, there will be sectors and companies that are unwilling or unable to make the necessary transition (The World Bank, 2021), hence the investment costs associated with making the transition will be different for high- and low-quality firms (i.e., signal cost; Bird & Smith, 2005; Bonifant et al., 1995). Indeed, assuming that sustainability initiatives are "honest signals," they can be considered more costly for low-quality firms, who require more investment to improve their ability to decarbonize, than for high-quality firms, as explained by Connelly et al. (2011, p. 45): "If a signaler does not have the underlying quality associated with the signal but believes the benefits of signaling outweigh the costs of producing the signal, the signaler may be motivated to attempt false signaling."

The truthfulness of the signal is a fundamental characteristic. This concept is defined as "signal honesty," which refers to the extent to which the underlying quality of the signal is present in the signaler (Connelly et al., 2011). If signals are not perceived as honest, it may lead to both signalers and signals having a dishonest reputation, which decreases the value of the signal perceived by investors (Cohen & Dean, 2005). For disclosure on current ESG performance, the truthfulness or validity of the signaled information is high since this disclosure typically has to comply with (regulatory) standards and is validated by third-party auditors. The signal honesty of ESG targets is less straightforward, with it being impossible to provide factual information about a firm's future ESG performance. Instead, the honest signaling depends on the current resources of the firm, including available human and intellectual capital and detailed decarbonization plans. However, the underlying quality of sustainability-related activities will only become clear to investors further in the future. Therefore, the extent to which early adopters are genuinely able to reach their respective targets will largely determine the reputation and signal honesty of the initiatives. From their analysis of early adopters, Giesekam et al. (2021) found that most targets were on track to be achieved, thus providing an early indication that signal honesty is adequate and that the target submission process accurately screens for companies with the necessary resources to achieve sustainability goals.

Two final concepts focus on whether signals are visible to receivers (signal observability), and whether receivers are looking for signals (receiver attention). First, signal observability is a particularly relevant concept, because "if actions insiders take are not readily observed by outsiders, it is difficult to use those actions to communicate with receivers" (Connelly et al., 2011, p. 45). Since firms typically disclose ESG activities in their reporting, approved targets are presented and significant media attention is typically given (Freiberg et al., 2021; Trexler & Schendler, 2015). Receiver attention is instead defined as the degree to which receivers are actively searching for the signal (Connelly et al., 2011). For investors, as described in the beginning of this section, ESG information is increasingly valuable, particularly when it is concise, comparable, and credible (Cohen et al., 2015). As such, considering the information it holds about the future of the firm and what it tells about the firm's resources, investors are expected to be vigilant for firms disclosing ESG information, particularly when they have been validated by a credible third-party. Signaling theory posits that effective signals require high observability and high attention. As such, the stage is set for the signaled information to influence the behavior of investors with regard to their analysis of firms' ESG information.

## *Value Relevance Theory*

As Barth et al. (2001, p. 95) described it, the value relevance theory examines "the association between accounting amounts and equity market values," or as put it by Beisland (2009, p. 8), "the usefulness of accounting information from the perspective of equity investors." This theory has often been used to see how financial as well as non-financial information can provide insights on firm value (Miralles-Quirós et al., 2018; De Klerk et al., 2015).

In his seminal paper, James Ohlson proposed a valuation framework which incorporates both fundamental financial indicators and the net surplus relation principle, which asserts that any alteration in book value should equate to earnings minus dividends (Ohlson, 1995, p. 661). Ohlson (1995) identifies that the value of a firm is influenced by a combination of financial and non-financial factors, by examining the correlation between a firm's market worth and its prospective earnings, book values, and dividends. According to this model, a firm's value is determined by a blend of capitalized earnings and book value.

This seminal work forms the basis of what later becomes known as the Ohlson price model. The model proposes that a company's market value is shaped not only by financial metrics but also by unspecified non-financial data, such as ESG disclosure.

Ohlson (1995, p. 663) identifies the following assumptions for the proposed valuation model:

1. The present value of expected dividends determines the market value.
2. The model satisfies the net surplus relation.
3. Dividends reduce book value without affecting current earnings.
4. A linear model frames the stochastic time-series behavior of abnormal earnings.

The linear model includes information of abnormal earnings and other information and places no restrictions on the variances and covariances of the disturbance terms (Ohlson, 1995, p. 668). The assumptions are used to formulate a "linear, closed-form, valuation solution explaining goodwill, that is, value equals book value plus a linear function of current abnormal earnings and the scalar variable representing other information" (Ohlson, 1995, p. 664).

From a financial standpoint, Collins et al. (1997) investigate the collective value relevance of earnings and book values, noting a rise in their combined relevance over the previous four decades. They observe a transition in value relevance from earnings towards book values, primarily attributed to exceptional one-time items and a higher occurrence of negative earnings (Collins et al., 1997, p. 40). This trend is further examined by Francis and Schipper (1999), who demonstrate a diminishing importance of earnings data alongside an augmented significance of balance sheet and book value information.

In a similar vein, Barth and Clinch (2009, p. 254) examine the relationship between equity market value, equity book value, and earnings by employing a modified version of the Ohlson (1995) valuation model. They focus on how five distinct scale-related effects could introduce causal inference challenges. These effects include multiplicative and additive correlated omitted scale variables, scale-varying valuation parameters, survivorship, and scale-related heteroskedasticity (Barth & Clinch, 2009, p. 253). The aim of their research is to enhance understanding of how scale effects may potentially result in erroneous conclusions. They explore various potential remedies for addressing scale effects and determine that specifications involving share prices and non-deflated market value of equity are the most effective (Barth & Clinch, 2009, p. 255).

## Review of the Empirical Literature

A growing body of empirical research has delved into the effects of non-financial disclosure on information asymmetry—typically measured in terms of forecast errors or dispersion—focusing on choice of reporting framework, namely sustainability reporting and integrated reporting. Within the realm of CSR disclosure, Plumlee et al. (2015) and Matsumura et al. (2014) explored disclosures pertaining to specific activities, such as environmental endeavors. Conversely, Dhaliwal et al. (2011, 2012) analyzed the impact of standalone CSR reports encompassing all CSR activities. They found that firms initiating CSR reports experienced lower costs of equity capital and reduced analyst forecast errors, particularly when supported by superior CSR activities. Similarly, Dhaliwal et al. (2012) observed a stronger impact of standalone sustainability reports on information asymmetry, particularly in more stakeholder-oriented countries.

In a niche investigation on climate-related physical risk reporting, Schiemann and Sakhel (2019) noted decreased information asymmetry for companies disclosing serious physical risks in high carbon-emitting sectors. Muslu et al. (2019) developed a scoring system based on ESG disclosure quality, observing that more substantial content in ESG reports led to lower forecast accuracy, indicating a positive association between CSR disclosure and forecast accuracy.

Zhou et al. (2017) and Bernardi and Stark (2018) found significantly lower forecast errors for companies adhering to the IR Framework, indicating the usefulness of IR for earnings forecasting. Caglio et al. (2020) noted a significant reduction in forecast dispersion for assured IR reports. However, studies in voluntary settings yielded mixed results. Kim et al. (2017) and Flores et al. (2019) demonstrated a decrease in forecast dispersion and improved forecast errors with voluntary IR adoption. By contrast, Wahl et al. (2020) found no evidence supporting a reduction in information asymmetry with voluntary IR disclosure. Lai et al. (2018) indicated that the impact of voluntary IR release on forecast accuracy is contingent upon institutional characteristics of countries. Only Permatasari and Narsa (2022) directly compared sustainability reporting (SR) and IR, finding that SR exhibits higher value relevance than IR.

Focusing on the market consequences of improved ESG performance and consequent decreased information asymmetry, Miralles-Quirós et al. (2018) examine the effect of CSR performance levels among companies listed on the São Paulo Stock Exchange over a six-year span from 2010 to 2015. Grounding their analytical approach in Ohlson's price model, they examined three distinct models: (1) assessing the correlation between financial indicators such as book value and earnings with stock price, (2) incorporating a measure of CSR performance to evaluate its relevance to stock price, and (3) introducing a variable for firms operating in environmentally sensitive sectors to test potential variations in CSR performance valuation across industries (Miralles-Quirós et al., 2018, p. 5). Their findings lend support to the value-enhancing theory over the shareholder expense theory, though also indicated that the market does not place significant value on the three ESG pillars.

De Klerk et al. (2015) explore the value relevance of CSR disclosure on share prices within the UK context. Incorporating the share price specification recommended by Barth and Clinch (2009), they investigate whether higher levels of CSR disclosure among the 100 largest UK companies are linked to elevated share prices, employing three measures of CSR disclosure: an indicator variable indicating whether the GRI framework is utilized; the degree of compliance with these guidelines; and finally, a composite measure of CSR disclosure practices derived from third-party data (De Klerk et al., 2015, p. 212). Their findings demonstrate a positive association between GRI-related disclosure levels and share prices. The authors conclude that there is "evidence that the level of CSR disclosure by UK companies is value relevant in such a way that higher levels of CSR disclosure are associated with higher share prices" (De Klerk et al., 2015, p. 210). They suggest that information asymmetry explains why higher CSR disclosure is positively correlated with share price. Moreover, they find supportive evidence for their hypothesis that greater CSR disclosure among companies operating in environmentally sensitive industries is linked to higher share prices (De Klerk et al., 2015, p. 224). Through CSR disclosure, often proxied by ESG performance, investors gain valuable insights for assessing companies' future cash flow risks.

In a similar vein, Kaspereit and Lopatta (2016) investigate whether the Sustainable Asset Management Group's corporate sustainability ranking and GRI sustainability reporting are value relevant and associated with higher market valuation among the 600 largest European listed companies over an eleven-year period from 2001 to 2011. Employing the Ohlson valuation model for their study, they assess the value relevance of being in the top quintile of a sector in terms of corporate sustainability, and secondly, analyze the value relevance of self-reported data according to GRI guidelines (Kaspereit & Lopatta, 2016, p. 2). By utilizing both self-reported data from company reports and the Sustainable Asset Management index ranking, they simultaneously explore the value relevance of third-party corporate sustainability metrics and companies' voluntary sustainability disclosure. Their findings reveal a positive correlation between corporate sustainability and market valuation, indicating that sustainability reporting is viewed positively by capital markets (Kaspereit & Lopatta, 2016, p. 18). However, they do not observe any impact of external assurance on the value relevance of GRI application levels, nor do they find evidence of an interaction or distinction between being sustainable and claiming to be sustainable.

If, from a theoretical standpoint, improved disclosure and therefore ESG performance should reduce information asymmetry (and consequently improve a company's market performance), the empirical evidence base shows conflicting results which have been in part attributed to sector-specific effects. Indeed, neo-institutional theory suggests that institutional changes follow a structured pattern, leading organizations to exhibit similar behaviors. This phenomenon, termed institutional isomorphism by DiMaggio and Powell (1991), encompasses mimetic, coercive, and normative isomorphism. In brief, mimetic isomorphism involves organizations imitating the behaviors of others, coercive isomorphism entails adapting to external pressures, and normative isomorphism involves adopting specific systems or standards to gain recognition. In line with neo-institutional theory, organizations' decisions to disclose their CSR activities are influenced by various environmental forces. These organizations respond to internal and external pressures by engaging in mimetic, coercive, or normative isomorphism. Thus, empirical investigations into the relationship between ESG factors and information asymmetry must consider variations in organizational behavior influenced by regulations, geographic location, industry, economic conditions, and other factors systematically shaping this relationship, all of which supports an industry-level empirical investigation focusing on the European context, such as that presented in the sections below.

## Research Methods

### *Forecast Accuracy*

In the empirical literature, information asymmetry has been most frequently measured through the accuracy of financial forecasts evaluated by the magnitude of the forecast error (FE). Consequently, the lower the error, the higher the accuracy of the forecast. Authors such as Zhou et al. (2017) estimated these forecast errors based on the absolute difference between the actual earnings per share (EPS) of a company obtained in a given year and the average financial analysts' forecasts for the same period, scaled by the company's average stock price, as follows:

$$FE_{i,t} = \frac{\left|EPS_{i,t} - \text{Forecast } EPS_{i,t}\right|}{P_{i,t}}$$

Where: EPS_(i,t) represents the actual earnings per share of company "i" in year "t," Forecast EPS_(i,t) represents the median forecast for company "i" in year "t," and P_(i,t) denotes the average stock price of company "i" in year "t."

### *Explanatory and Control Factors*

As with Chap. 5, the ESG and all the other relevant company data used for the analysis presented sourced and downloaded from the Refinitiv platform (Eikon, 2022). With Refinitiv, the ESG scores are formulated to impartially assess a company's ESG performance, dedication, and efficacy using data provided by the company. The three main ESG pillars (environmental, social, and governance) comprise ten ESG categories, namely, resource use, emissions reduction, innovation, workforce, human rights, community, product responsibility, management, shareholders, and CSR strategy score. Refinitiv also provides a comprehensive corrected ESG score that takes into account any significant ESG-related issues the company may have faced.

To enable a quantitative evaluation of the relationship between ESG performance and information asymmetry, data and information spanning the decade from 2012 to 2022 were extracted using the Datastream application for Microsoft Excel (Eikon, 2022). Following the retrieval of data from the platform (comprising 550 observations per company per year), they were processed to facilitate subsequent statistical analysis. Variables were chosen based on existing studies in the literature regarding the correlation between sustainability and economic performance.

The sample of analyzed health care companies belonged to the Euro STOXX 600 index. This is a stock index consisting of 600 public companies listed on capital markets across the European continent. It comprises a fixed number of 600 components, including large companies with significant market capitalizations from 17 European countries. This composition represents approximately 90% of the total market capitalization of the European stock market. By focusing on a single, albeit complex, geographical region, the research aims to reduce the risk of omitted-variable bias, a significant challenge in studies spanning multiple heterogeneous regions with varying financial reporting regulations, jurisdictions, and practices. This methodological approach allows for the simultaneous control of all fixed factors influencing both dependent and independent variables (De Jong et al., 2008). The countries included in the index are Portugal, the United Kingdom, Austria, Belgium, Germany, Denmark, Finland, France, Ireland, Italy, Norway, the Netherlands, Poland, Spain, Luxembourg, Sweden, and Switzerland. The index is classified into 11 sectors, with the health care sector representing 9.2% of the total, with 55 listed companies. In this regard, the health care sector holds a nearly average position in terms of relative proportion to the number of companies comprising the index.

The primary set of explanatory variables were ESG scores: total, environmental, social, and governance pillars, as well as the ten score categories—which were treated as continuous variables. To account for determinants of information asymmetry, contextual and financial controls were included based on prior published research which has been reviewed and presented in the respective section above. These controls comprise several factors, namely:

- Country of domicile.
- Size, determined by metrics such as market value of equity, total assets, total revenues, market capitalization, and number of employees at the end of the fiscal year (all natural logarithms, ratio scale variable).
- Leverage, computed as the ratio of total debt to total assets at the end of the fiscal year.

- Beta, indicating the volatility of the company's share compared to the broader market.
- Book value per share, reflecting the equity available to common shareholders relative to the number of shares outstanding.
- Market-to-book value, evaluating the potential over or undervaluation of a company's share by comparing its market price with net assets.

### *Statistical Analysis*

Summary statistics were initially utilized to depict the characteristics of the company sample and explore relationships between selected variables. Pearson's pairwise correlations were then computed to evaluate the direction and strength of univariate linear associations. Linear regression analyses followed a forward stepwise approach, progressively constructing models to investigate the relationship between ESG scores and forecast error. In these models, β coefficients represented the regression coefficient of interest, while controlling for relevant factors (both contextual and financial controls).

Specification 1 included only the primary explanatory variable, namely, the total ESG score. Subsequent specifications incorporated the scores for the three ESG pillars (Model 2) and then all selected controls (Model 3). To compare predictor effects across multiple regression frameworks, all identified variables were retained in the final model specifications. Interaction terms were tested to assess potential effect modifications, and Wald tests were performed to assess differences between categories. Statistical significance was determined at a $p$-value $<0.05$ level. The analyses were conducted using STATA 16 software (StataCorp, 2019).

## Relationship Between ESG and Information Asymmetry

Figure 6.1 depicts the dispersion of forecast errors, expressed in percentage terms for the European health care industry. It reveals that for the majority of companies, the average forecast error falls within the range of 0.00–5.00%. However, for a small portion of the sample, this error can extend up to 15% on average.

From a quantile distribution perspective, Fig. 6.2 confirms the findings presented in the previous figure. For 25% of the sample, forecast accuracy appears to be optimal, while for 5% of the sample, errors are significantly pronounced, surpassing the 5% threshold and reaching 15% in the most extreme cases.

From a computational standpoint, however, the median estimate value is influenced by the number of analysts who have conducted the analysis and consequently by the evaluations of the listed company. In this regard, Fig. 6.3 depicts the distribution of the number of estimates made, revealing that for most cases, publicly

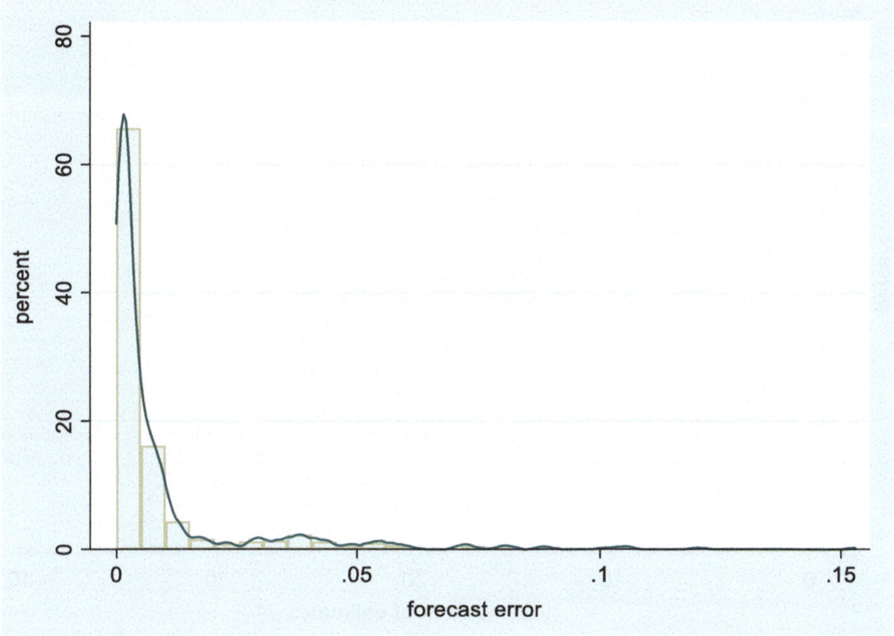


**Fig. 6.1** Percentage distribution of forecast errors

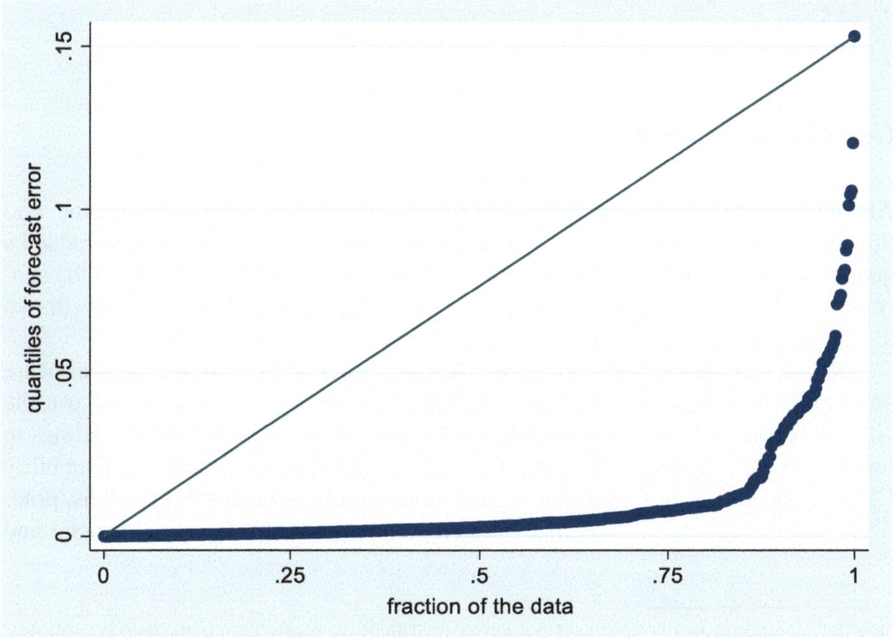


**Fig. 6.2** Quantile distribution of forecast errors

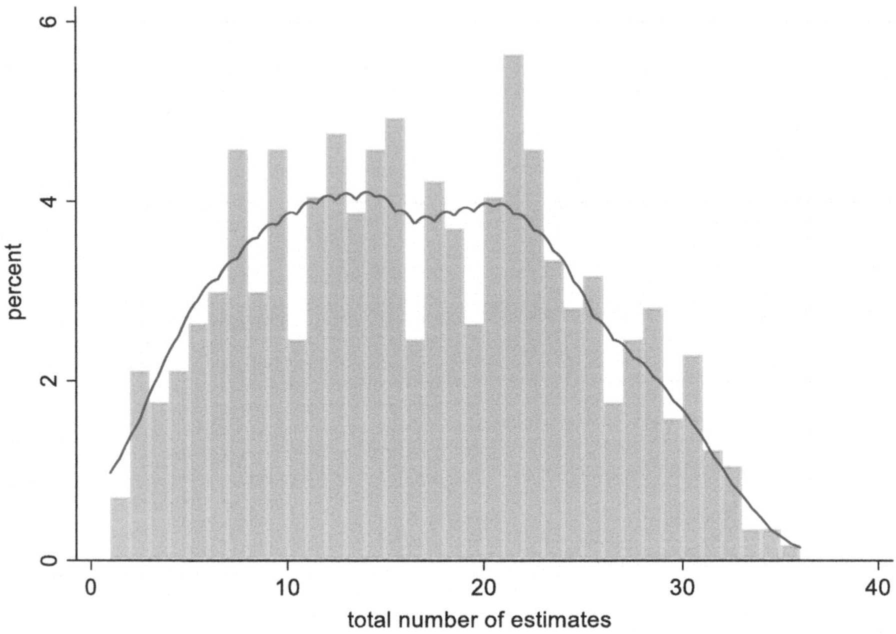


**Fig. 6.3** Distribution of the number of estimates made

traded healthcare companies are followed by 10 to 30 analysts, with a range from 2 to 35.

## *Correlation Analysis*

An initial exploratory analysis employing Pearson's linear coefficient and Spearman's coefficient for potentially non-linear monotonic relationships enabled a quantitative assessment of the strength and direction of associations among the analyzed variables.[2] Technically, these coefficients are computed based on a line drawn to minimize the distance between observed points.

Table 6.1 compares forecast errors (fe) with three choices made by healthcare enterprises regarding non-financial reporting: gri—whether they opted to use the Global Reporting Initiative framework or at least followed its general guidelines in report compilation; csrsr—whether the company publishes a separate sustainability CSR report or allocates a section of the annual report to sustainability plans, practices, and impact; csria—whether the company explicitly integrates financial and

[2] For the operational details employed in the statistical analysis, please refer to the text: Wooldridge, J. M. (2012). Introductory econometrics: A modern approach. Mason, OH: South-Western Cengage Learning.

**Table 6.1** Reporting choice and forecast accuracy

| Variables | (1) | (2) | (3) | (4) |
|---|---|---|---|---|
| (1) fe | 1.000 | | | |
| (2) gri | −0.056 | 1.000 | | |
| | (0.353) | | | |
| (3) csrsr | −0.023 | −0.035 | 1.000 | |
| | (0.624) | (0.596) | | |
| (4) csria | −0.025 | −0.203* | 0.134 | 1.000 |
| | (0.691) | (0.014) | (0.058) | |

***$p < 0.01$, **$p < 0.05$, *$p < 0.1$. Notes: variable definitions available in Appendix Variables selected for analysis

**Table 6.2** Total and environmental score and forecast accuracy

| Variables | (1) | (2) | (3) | (4) | (5) | (6) | (7) |
|---|---|---|---|---|---|---|---|
| (1) fe | 1.000 | | | | | | |
| (2) epsy1 | −0.036 | 1.000 | | | | | |
| (3) esgscore | 0.018 | 0.530* | 1.000 | | | | |
| (4) envscore | −0.119* | 0.561* | 0.793* | 1.000 | | | |
| (5) resuse | −0.097* | 0.606* | 0.778* | 0.872* | 1.000 | | |
| (6) emiss | −0.110* | 0.393* | 0.661* | 0.860* | 0.686* | 1.000 | |
| (7) envinn | −0.087 | 0.360* | 0.475* | 0.669* | 0.482* | 0.433* | 1.000 |

***$p < 0.01$, **$p < 0.05$, *$p < 0.1$. Notes: variable definitions available in Appendix Variables selected for analysis

**Table 6.3** Social scores and forecast accuracy

| Variables | (1) | (2) | (3) | (4) | (5) | (6) |
|---|---|---|---|---|---|---|
| (1) fe | 1.000 | | | | | |
| (2) socscore | 0.010 | 1.000 | | | | |
| (3) workforce | −0.078 | 0.643* | 1.000 | | | |
| (4) humrights | 0.007 | 0.732* | 0.314* | 1.000 | | |
| (5) commun | 0.018 | 0.804* | 0.413* | 0.520* | 1.000 | |
| (6) prodresp | 0.025 | 0.722* | 0.442* | 0.282* | 0.396* | 1.000 |

***$p < 0.01$, **$p < 0.05$, *$p < 0.1$. Notes: variable definitions available in Appendix Variables selected for analysis

non-financial factors in the management and analysis section within the annual financial report. From the table, it can be deduced that, from a univariate perspective for publicly listed companies in Europe within the healthcare sector, there is no significant relationship with forecast accuracy. Furthermore, it is notable that none of the considered non-financial reporting choices seem to have, on average, any influence on analysts' ability to predict companies' financial performance. However, it is observed that the choice to use the Global Reporting Initiative framework is significantly and negatively associated with a section dedicated to sustainability in the annual report.

**Table 6.4** Governance scores and forecast accuracy

| Variables | (1) | (2) | (3) | (4) | (5) | (6) |
|---|---|---|---|---|---|---|
| (1) fe | 1.000 | | | | | |
| (2) govscore | 0.126* | 1.000 | | | | |
| (3) manag | 0.126* | 0.956* | 1.000 | | | |
| (4) sharehold | 0.100* | 0.525* | 0.311* | 1.000 | | |
| (5) csrstrat | −0.015 | 0.484* | 0.354* | 0.052 | 1.000 | |
| (6) esgcontr | −0.049 | −0.332* | −0.322* | −0.013 | −0.374* | 1.000 |

***$p < 0.01$, **$p < 0.05$, *$p < 0.1$. Notes: variable definitions available in Appendix Variables selected for analysis

Tables 6.2, 6.3, and 6.4, on the other hand, focus on the relationship between forecast errors and the three ESG pillars and their respective 10 score categories. In particular, Table 6.2 compares estimation accuracy with the number of estimates made (epsy1), the total ESG score provided by Refinitiv (esgscore), and the environmental score (envscore) with their respective belonging categories related to environmental score: resuse—company's ability to efficiently utilize necessary resources, emiss—the company's commitment and ability to minimize harmful emissions from production and operational processes; envinn—the company's capacity to innovate through the generation of environmental technologies and eco-friendly processes.

The table illustrates how the recent set of values pertaining to environmental impact is inversely correlated with forecast errors in financial performance, consequently leading to enhanced estimation accuracy by analysts.

Table 6.3 shows evidence for a significant relationship with the social score (socscore) and each of its four categories of component scores, namely: workforce—which assesses a company's ability to satisfy employees in the workplace environment; human rights—which evaluates the company's respect for fundamental individual rights; commun—measuring the company's commitment to public health protection and business ethics; prodresp—assessing the company's production capacity while adhering to principles of integrity, safety, and privacy. None of these dimensions yielded significant results in terms of their effect on estimation accuracy.

Lastly, Table 6.4 illustrates the aforementioned relationship with corporate governance variables. This includes both the total score (govscore) and aspects such as the company's ability to adhere to corporate governance principles (manag), impartial treatment of shareholders (sharehold), communication strategy regarding the integration of value associated with the three ESG pillars (csrstrat), and the company's media exposure concerning controversies related to the three sustainability factors. In this regard, positive and significant relationships emerge between forecast errors and the total corporate governance score, corporate governance practices, and shareholder treatment. It is also noteworthy to observe a very high correlation (0.956) between the management and total governance scores, indicating the mutual influence of these scores.

## *Regression Analysis*

To predict how a change in ESG scores affects forecast error, a series of regression analyses were conducted. Beginning with a univariate analysis between the total score, the scores of the three ESG pillars, and the score related to ESG controversies, Table 6.5 presents the estimated linear regression coefficients. In summary, concerning the healthcare sector over the past decade, the score related to environmental performance (mean −0.000031, standard error (0.000012)) and the corporate governance score (mean 0.000037, standard error (0.000014)) appear to play a significant role in influencing estimation accuracy.

Specifically, with a comparable effect magnitude, an increase in the environmental score appears to reduce forecast error and consequently enhance analysts' forecast accuracy, while a better corporate governance score is, on average, associated with increased forecast error. However, the other scores do not seem to play a role in this regard.

Table 6.6 delves into the univariate associations between the scores of the ten ESG scoring categories belonging to the three pillars and the accuracy of estimates.

**Table 6.5** Univariate linear association between ESG and forecast error

| Variable | Total ESG | Environmental | Social | Governance | ESG controversies |
|---|---|---|---|---|---|
| Coefficient | 0.000007 | −0.000031** | 0.000003 | 0.000037*** | −0.000012 |
| | (0.000018) | (0.000012) | (0.000016) | (0.000014) | (0.000012) |
| Constant | 0.003929*** | 0.006186*** | 0.004123*** | 0.002024** | 0.005466*** |
| | (0.001231) | (0.000765) | (0.001226) | (0.000923) | (0.001079) |
| Observations | 456 | 456 | 456 | 456 | 456 |
| R-squared | 0.000318 | 0.014185 | 0.000105 | 0.015790 | 0.002396 |

Standard errors in parentheses ***$p < 0.01$, **$p < 0.05$, *$p < 0.1$. Notes: variable definitions available in Appendix Variables selected for analysis

**Table 6.6** Univariate linear association between ESG score categories and forecast error

| Variable | β | SE | constant | SE | $R^2$ |
|---|---|---|---|---|---|
| resource use | −0.000022** | (0.000011) | 0.005928*** | (0.000803) | 0.00937 |
| emissions | −0.000026** | (0.000011) | 0.006056*** | (0.000771) | 0.01204 |
| env. Innov. | −0.000019* | (0.000010) | 0.004946*** | (0.000424) | 0.00759 |
| workforce | −0.000032* | (0.000019) | 0.007016*** | (0.001616) | 0.00601 |
| human rights | 0.000001 | (0.000009) | 0.004289*** | (0.000681) | 0.00005 |
| comm. | 0.000004 | (0.000011) | 0.004092*** | (0.000808) | 0.00033 |
| prod. Resp. | 0.000006 | (0.000012) | 0.003894*** | (0.000966) | 0.00062 |
| management | 0.000031*** | (0.000011) | 0.002315*** | (0.000818) | 0.01592 |
| shareholders | 0.000022** | (0.000010) | 0.003129*** | (0.000659) | 0.00991 |
| CSR strategy | −0.000003 | (0.000010) | 0.004565*** | (0.000632) | 0.00023 |

Standard errors in parentheses ***$p < 0.01$, **$p < 0.05$, *$p < 0.1$. Notes: variable definitions available in Appendix Variables selected for analysis

Concerning the categories belonging to the environmental score, namely the ability to use resources efficiently, the commitment and ability of the enterprise to reduce harmful emissions, and innovation through the generation of eco-friendly technologies, all indicate a prediction of reduced forecast error. Regarding the social score, partly reflecting what was observed in Table 6.3, the company's ability to satisfy employees in their workplace appears to be the only category influencing information asymmetry. However, some inconsistency is noted for the categories belonging to pillar G, where the company's ability to adhere to corporate governance principles and impartial treatment of shareholders are positively associated with forecast error, while the communication strategy regarding the integration of value associated with the three ESG pillars negatively influences it, albeit to a lesser extent compared to other scoring categories.

Moving on to the multivariate associations between forecast error and ESG scores, Table 6.7 presents the estimates calculated on the sample of 600 European companies examined, aiming to correct the univariate associations mentioned earlier for the effects of other factors, both fixed and time-varying, that may affect analysts' ability to predict the financial performance of listed companies in the healthcare sector. While the total ESG score did not seem to have predictive capability for forecast error under univariate analysis (Model 1), this effect was actually masked by other factors, particularly the return on assets and the number of analysts following the company, both of which negatively influence estimation accuracy. Conversely, the book value per share (BVPS), i.e., the net asset value per share, and

**Table 6.7** Multivariate association between ESG score and forecast error

| Variables | Model 1 | Model 2 | Model 3 | Model 4 |
|---|---|---|---|---|
| ESG score | 0.000007 | 0.000045** | 0.000038* | 0.000054** |
| | (0.000018) | (0.000020) | (0.000022) | (0.000022) |
| epsy1 | | −0.000134*** | −0.000121*** | −0.000115*** |
| | | (0.000044) | (0.000043) | (0.000043) |
| currency | | 0.000303*** | 0.000212* | 0.000151 |
| | | (0.000108) | (0.000111) | (0.000109) |
| brdnonex | | | −0.000061*** | −0.000050** |
| | | | (0.000023) | (0.000024) |
| brdind | | | 0.000030** | 0.000020 |
| | | | (0.000014) | (0.000014) |
| roa | | | | −0.000138*** |
| | | | | (0.000034) |
| bvps | | | | −0.000037** |
| | | | | (0.000014) |
| Constant | 0.003929*** | 0.002451** | 0.006393*** | 0.007089*** |
| | (0.001231) | (0.001231) | (0.002084) | (0.002151) |
| Observations | 456 | 455 | 455 | 455 |
| R-squared | 0.000318 | 0.048754 | 0.070109 | 0.109069 |

Standard errors in parentheses ***$p < 0.01$, **$p < 0.05$, *$p < 0.1$. Notes: variable definitions available in Appendix Variables selected for analysis

**Table 6.8** Multivariate association between E score and forecast error

| Variables | Model 1 | Model 2 | Model 3 | Model 4 |
|---|---|---|---|---|
| E score | −0.000031** | −0.000006 | −0.000008 | 0.000003 |
| | (0.000012) | (0.000014) | (0.000014) | (0.000014) |
| epsy1 | | −0.000071 | −0.000071 | −0.000067 |
| | | (0.000044) | (0.000044) | (0.000043) |
| currency | | 0.000329*** | 0.000221** | 0.000178 |
| | | (0.000108) | (0.000111) | (0.000110) |
| brdnonex | | | −0.000053** | −0.000045* |
| | | | (0.000023) | (0.000024) |
| brdind | | | 0.000040*** | 0.000032** |
| | | | (0.000013) | (0.000013) |
| roa | | | | −0.000130*** |
| | | | | (0.000035) |
| bvps | | | | −0.000031** |
| | | | | (0.000014) |
| Constant | 0.006186*** | 0.004602*** | 0.007199*** | 0.008188*** |
| | (0.000765) | (0.000976) | (0.002051) | (0.002121) |
| R-squared | 0.014185 | 0.038301 | 0.064467 | 0.096988 |

Standard errors in parentheses ***$p < 0.01$, **$p < 0.05$, *$p < 0.1$; Notes: variable definitions available in Appendix Variables selected for analysis

the proportion of non-executive members on the board of directors appear to be negatively associated with forecast error.

Now turning to the analysis of ESG scores related to its component pillars, Table 6.8 explores the multivariate association with the environmental factor. Contrary to what was observed for the total score, once adjusted for the effect of control variables, the environmental score does not appear to be significantly associated with forecast accuracy. However, the effects of the selected control variables remain comparable, predicting the dependent variable with signs and magnitudes of effect similar to those observed for Table 6.7, except for the number of actual financial performance estimates, which does not seem to play a significant role.

In contrast to the previous table, Table 6.9 indicates that the social score is not significant in univariate analysis but becomes influential on forecast accuracy as the effect of control variables is taken into account. Additionally, the control variables align in direction and magnitude of effect with those estimated for the total score, except for the proportion of independent members on the board of directors for which an increase predicts a deterioration in forecast accuracy.

Regarding the G score for corporate governance in Table 6.10, results comparable to those observed for the social score emerge for the control variables. However, the correction pattern across the four specified models differs, remaining consistent with the increase in the number of variables, with a negative effect on information asymmetry.

**Table 6.9** Multivariate association between S score and forecast error

| Variables | Model 1 | Model 2 | Model 3 | Model 4 |
|---|---|---|---|---|
| S score | 0.000003 | 0.000031* | 0.000022 | 0.000036* |
| | (0.000016) | (0.000017) | (0.000018) | (0.000018) |
| epsy1 | | −0.000118*** | −0.000106** | −0.000099** |
| | | (0.000042) | (0.000042) | (0.000042) |
| currency | | 0.000330*** | 0.000232** | 0.000180 |
| | | (0.000107) | (0.000111) | (0.000109) |
| brdnonex | | | −0.000057** | −0.000047** |
| | | | (0.000023) | (0.000024) |
| brdind | | | 0.000034** | 0.000024* |
| | | | (0.000013) | (0.000013) |
| roa | | | | −0.000138*** |
| | | | | (0.000034) |
| bvps | | | | −0.000035** |
| | | | | (0.000014) |
| Constant | 0.004123*** | 0.002774** | 0.006404*** | 0.007044*** |
| | (0.001226) | (0.001266) | (0.002128) | (0.002190) |
| R-squared | 0.000105 | 0.044790 | 0.066923 | 0.104640 |

Standard errors in parentheses ***$p < 0.01$, **$p < 0.05$, *$p < 0.1$; Notes: variable definitions available in Appendix Variables selected for analysis

**Table 6.10** Multivariate association between G score and forecast error

| Variables | Model 1 | Model 2 | Model 3 | Model 4 |
|---|---|---|---|---|
| G score | 0.000037*** | 0.000046*** | 0.000044*** | 0.000045*** |
| | (0.000014) | (0.000014) | (0.000015) | (0.000015) |
| epsy1 | | −0.000118*** | −0.000107*** | −0.000088** |
| | | (0.000038) | (0.000038) | (0.000038) |
| currency | | 0.000224** | 0.000141 | 0.000089 |
| | | (0.000111) | (0.000114) | (0.000112) |
| brdnonex | | | −0.000064*** | −0.000051** |
| | | | (0.000023) | (0.000024) |
| brdind | | | 0.000025* | 0.000020 |
| | | | (0.000013) | (0.000013) |
| roa | | | | −0.000124*** |
| | | | | (0.000034) |
| bvps | | | | −0.000036** |
| | | | | (0.000014) |
| Constant | 0.002024** | 0.002533** | 0.006755*** | 0.007566*** |
| | (0.000923) | (0.001032) | (0.002032) | (0.002107) |
| R-squared | 0.015790 | 0.061162 | 0.081298 | 0.114930 |

Standard errors in parentheses ***$p < 0.01$, **$p < 0.05$, *$p < 0.1$; Notes: variable definitions available in Appendix Variables selected for analysis

**Table 6.11** Multivariate association between ESG score categories and forecast error

| Variables | Coefficient |
|---|---|
| Resource use | −0.000034* |
| | (0.000019) |
| Emissions | −0.000021 |
| | (0.000016) |
| Environmental innovation | −0.000010 |
| | (0.000012) |
| Workforce | −0.000026 |
| | (0.000024) |
| Human rights | 0.000018 |
| | (0.000013) |
| Community | 0.000020 |
| | (0.000015) |
| Product responsibility | 0.000021 |
| | (0.000014) |
| Management | 0.000032** |
| | (0.000013) |
| Shareholders | 0.000009 |
| | (0.000011) |
| CSR strategy | −0.000004 |
| | (0.000014) |
| Constant | 0.003914** |
| | (0.001789) |
| Observations | 456 |
| R-squared | 0.058220 |

Standard errors in parentheses ***$p < 0.01$, **$p < 0.05$, *$p < 0.1$; Notes: variable definitions available in Appendix Variables selected for analysis

Finally, for completeness, Table 6.11 displays the independent effect that each of the ten scoring categories comprising the overall ESG score has on the aforementioned accuracy, after adjusting for the influence of others. This reciprocal adjustment demonstrates how only the dimensions of efficient resource usage and the company's ability to adhere to corporate governance principles can significantly predict overall forecast error, and in opposite directions.

## Concluding Remarks

This chapter has addressed the relationship between the accuracy of financial analysts' forecasts, non-financial reporting choices, and ESG scores. The presented results indicate that various choices related to the non-financial reporting framework, communication of social responsibility activities, and activities and composition of listed companies significantly influence, in a heterogeneous manner, the ability of analysts to assess and estimate the financial performance expectations of companies, thereby impacting information asymmetry between capital and stock markets and information intermediaries. Placing these results within the context of published literature, a limited body of studies has investigated the relationship between forecast errors and non-financial reporting choices. Previous studies have focused particularly on the choice of framework, notably between separate sustainability reporting and the use of integrated reporting. In general, most of these studies indicate that the choice of the former type of framework positively, and to a greater extent, influences information asymmetry compared to other options available to companies (Higgins and Walker, 2012; Rowbottom and Locke, 2016; Lai et al., 2018).

More recently, a study conducted by Permatasari and Narsa (2022), comparing the relevance for investors of the sustainability report value versus integrated reporting, indicated a higher relevance for the latter type of framework. A study recently conducted by Rossi and Candio (2023) within the sample of the top 600 companies listed on European stock exchanges defined no significant differences in terms of the effect on information asymmetry between these two reporting options, but both were significantly superior in including sustainability information within the annual financial statements' notes. The strengths of this study lie in its focus on the healthcare sector and listed companies in Europe, where essential data for analysis are available, allowing for replicable analyses. However, there are limitations to consider, including basing estimates on binary or essentially quantitative variables.

The quality of the information contained, in addition to the chosen framework type and the conduct of CSR activities, can influence analysts' financial forecasting ability. Furthermore, this study focused on Europe, where there is a voluntary regime in terms of choosing the framework to use, unlike other contexts such as, for example, the Republic of South Africa where the use of integrated reporting is mandatory (Dumitru & Dragomir, 2021). Additionally, differences in legislative and regulatory accounting contexts can significantly affect the interpretability of the presented results, especially considering the healthcare sector, where, in contexts such as the United States, health care service coverage and demand vary in terms of both supply (private sector more prevalent than public) and demand—an older population demands proportionally more and different health care.

## References

Akerlof, G. (1970). The market for "lemons": Quality uncertainty and the market mechanism. *The Quarterly Journal of Economics, 84*(3), 488–500.

Barth, M. E., Beaver, W. H., & Landsman, W. R. (2001). The relevance of the value relevance literature for financial accounting standard setting: Another view. *Journal of Accounting and Economics, 31*(1–3), 77–104. https://doi.org/10.1016/S0165-4101(01)00019-2

Barth, M. E., & Clinch, G. (2009). Scale effects in capital markets-based accounting research. *Journal of Business Finance & Accounting, 36*(3–4), 253–288. https://doi.org/10.1111/j.1468-5957.2009.02133.x

Beisland, L. (2009). A review of the value relevance literature. *The Open Business Journal, 2*, 7–27. https://doi.org/10.2174/1874915100902010007

Bernardi, C., & Stark, A. W. (2018). Environmental, social and governance disclosure, integrated reporting, and the accuracy of analyst forecasts. *The British Accounting Review, 50*(1), 16–31.

Bird, R. B., & Smith, E. A. (2005). Signaling theory, strategic interaction, and symbolic capital. *Current Anthropology, 46*(2), 221–248.

Bonifant, B. C., Arnold, M. B., & Long, F. J. (1995). Gaining competitive advantage through environmental investments. *Business Horizons, 38*(4), 37–47.

Caglio, A., Melloni, G., & Perego, P. (2020). Informational content and assurance of textual disclosures: Evidence on integrated reporting. *European Accounting Review, 29*(1), 55–83.

Cohen, B. D., & Dean, T. J. (2005). Information asymmetry and investor valuation of IPOs: Top management team legitimacy as a capital market signal. *Strategic Management Journal, 26*(7), 683–690. https://www.jstor.org/stable/20142258

Cohen, J., Holder-Webb, L., & Zamora, V. (2015). Nonfinancial information preferences of professional investors. *Behavioral Research in Accounting, 27*(2), 127–153.

Collins, D. W., Maydew, E. L., & Weiss, I. S. (1997). Changes in the value-relevance of earnings and book values over the past forty years. *Journal of Accounting and Economics, 24*(1), 39–67. https://doi.org/10.1016/S0165-4101(97)00015-3

Connelly, B., Certo, S., Ireland, R., & Reutzel, C. (2011). Signaling theory: A review and assessment. *Journal of Management, 37*(1), 39–67.

De Jong, A., Kabir, R., & Nguyen, T. T. (2008). Capital structure around the world: The roles of firm- and country-specific determinants. *Journal of Banking & Finance, 32*(9), 1315–1344.

De Klerk, M., De Villiers, C., & Van Staden, C. (2015). The influence of corporate social responsibility disclosure on share prices: Evidence from the United Kingdom. *Pacific Accounting Review, 27*(2), 208–228. https://doi.org/10.1108/PAR-05-2013-0047

Dhaliwal, D. S., Li, Z., Tsang, A., & Yang, Y. (2011). Voluntary nonfinancial disclosure and the cost of equity capital: The initiation of corporate social responsibility reporting. *The Accounting Review, 86*(1), 59–100.

Dhaliwal, D. S., Radhakrishnan, S., Tsang, A., & Yang, Y. (2012). Nonfinancial disclosure and analyst forecast accuracy: International evidence on corporate social responsibility disclosure. *The Accounting Review, 87*(3), 723–759.

DiMaggio, P. J., & Powell, W. W. (1991). *The new institutionalism in organizational analysis*. The University of Chicago Press.

Dumitru, M., & Dragomir, V. D. (2021). The factors of integrated reporting quality: A meta-analysis. *SSRN Electronic Journal*. https://doi.org/10.2139/ssrn.3940094.

Eikon. (2022). *Environmental, social and governance scores from Refinitiv*. Retrieved from https://www.refinitiv.com/content/dam/marketing/en_us/documents/methodology/refinitiv-esg-scores-methodology.pdf

Elitzur, R., & Gavious, A. (2003). Contracting, signaling, and moral hazard: A model of entrepreneurs, 'angels,' and venture capitalists. *Journal of Business Venturing, 18*(6), 709–725.

Flores, E., Fasan, M., Mendes-da-Silva, W., & Sampaio, J. O. (2019). Integrated reporting and capital markets in an international setting: The role of financial analysts. *Business Strategy and the Environment, 28*(7), 1465–1480.

Flower, J. (2015). The international integrated reporting council: A story of failure. *Critical Perspectives on Accounting, 27*, 1–17. https://doi.org/10.1016/j.cpa.2014.07.002

Francis, J., & Schipper, K. (1999). Have financial statements lost their relevance. *Journal of Accounting Research, 37*(2), 319–352. https://www.jstor.org/stable/2491412

Freiberg, D., Grewal, J., & Serafeim, G. (2021). Science-based carbon emissions targets. *SSRN Electronic Journal*, 1–59.

García-Sánchez, I. M., & Noguera-Gámez, L. (2017). Integrated reporting and stakeholder engagement: The effect on information asymmetry. *Corporate Social Responsibility and Environmental Management, 24*(5), 395–413.
Gershon, R. A. (2013). Innovation failure: A case study analysis of Eastman Kodak and Blockbuster Inc. In Albarran, A. B. (Ed.), Media management and economics research in a transmedia environment, 46–68.
Giesekam, J., Norman, J., Garvey, A., & Betts-Davies, S. (2021). Science-based targets: On target? *Sustainability (Switzerland), 13*(4), 1–20. Global Sustainable Investment Alliance (GSIA).
Grossman, S. J., & Hart, O. D. (1981). The allocational role of takeover bids in situations of asymmetric information. *The Journal of Finance, 36*(2), 253–270.
Higgins, C., & Walker, R. (2012). Ethos, logos, pathos: Strategies of persuasion in social/environmental reports. *Accounting Forum, 36*(3), 194–208.
Kaspereit, T., & Lopatta, K. (2016). The value relevance of SAM's corporate sustainability ranking and GRI sustainability reporting in the European stock markets. *Business Ethics: A European Review, 25*(1), 1–24. https://doi.org/10.1111/beer.12079
Kim, S., Maas, K., & Perego, P. (2017). The effect of publication, format and content of integrated reports on analysts' earnings forecasts. In S. Boubaker, D. Cummings, & D. Nguyen (Eds.), *The handbook of finance and sustainability*. Edward Elgar. https://ssrn.com/abstract=2902549
Kirmani, A., & Rao, A. (2000). No pain, no gain: A critical review of the literature on signaling unobservable product quality. *Journal of Marketing, 64*(2), 66–79.
Lai, A., Melloni, G., & Stacchezzini, R. (2018). Integrated reporting and narrative accountability: The role of preparers. *Accounting, Auditing & Accountability Journal, 31*(5), 1381–1405.
Lambert, R. A., Leuz, C., & Verrecchia, R. E. (2012). Information asymmetry, information precision, and the cost of capital. *Review of Finance, 16*(1), 1–29.
Matsumura, E. M., Prakash, R., & Vera-Munoz, S. C. (2014). Firm-value effects of carbon emissions and carbon disclosures. *The Accounting Review, 89*(2), 695–724.
Miralles-Quirós, M. M., Miralles-Quirós, J. L., & Valente Gonçalves, L. M. (2018). The value relevance of environmental, social, and governance performance: The Brazilian case. *Sustainability, 10*(3), 574. https://doi.org/10.3390/su10030574
Muslu, V., Mutlu, S., Radhakrishnan, S., & Tsang, A. (2019). Corporate social responsibility report narratives and analyst forecast accuracy. *Journal of Business Ethics, 154*(4), 1119–1142.
Ohlson, J. A. (1995). Earnings, book values, and dividends in equity valuation. *Contemporary Accounting Research, 11*(2), 661–687. https://doi.org/10.1111/j.1911-3846.1995.tb00461.x
Permatasari, I., & Narsa, I. M. (2022). Sustainability reporting or integrated reporting: Which one is valuable for investors? *Journal of Accounting and Organizational Change, 18*(5), 666–684.
Plumlee, M., Brown, D., Hayes, R. M., & Marshall, R. S. (2015). Voluntary environmental disclosure quality and firm value: Further evidence. *Journal of Accounting and Public Policy, 34*, 336–361.
Rogers, E. M. (1995). *Diffusion of innovations* (4th ed.). The Free Press.
Rossi, P. (2019). *Beni immateriali, asimmetrie informative e analisti finanziari: Profili teorici, metodologie ed evidenze empiriche* – Milano: Franco Angeli, 2019 – 110 p. – Economia – Ricerche – ISBN: 9788891774972.
Rossi, P., & Candio, P. (2023). Sustainability reporting or integrated reporting: Which one is valuable for investors? *Journal of Accounting and Organizational Change, 18*(5), 666–684.
Rowbottom, N., & Locke, J. (2016). The emergence of <IR>. *Accounting and Business Research, 46*(1), 83–115.
Schiemann, F., & Sakhel, A. (2019). Carbon disclosure, contextual factors, and information asymmetry: The case of physical risk reporting. *European Accounting Review, 28*(4), 791–818.
Schoenmaker, D., & Schramade, W. (June 4, 2023). *Corporate finance for long-term value*. Springer. Available at SSRN: https://ssrn.com/abstract=4468886 or https://doi.org/10.2139/ssrn.4468886.
Schumpeter, J. A. (1942). *Capitalism, socialism and democracy* (Vol. 36, p. 132145). Harper & Row.
Spence, M. (1973). Job market signaling. *The Quarterly Journal of Economics, 87*(3), 355–374.

Spence, M. (2002). Signaling in retrospect and the informational structure of markets. *The American Economic Review, 92*(3), 434–459.

StataCorp. (2019). *Stata statistical software: Release 16*. College Station, TX: StataCorp LLC [program].

Stiglitz, J. (2001). Information and the change in the paradigm in economics. *The American Economic Review, 92*(3), 460–501.

Stubbs, W., & Higgins, C. (2018). Stakeholders' perspectives on the role of regulatory reform in integrated reporting. *Journal of Business Ethics, 147*, 489–508.

The World Bank. (2021). *Firm-level barriers on the path to net zero: How firms can make or break the green transition*. Retrieved from https://www.worldbank.org/en/events/2021/06/16/firm-levelbarriers-on-the-path-to-net-zero-how-firms-can-make-or-break-the-green-transition#1

Trexler, M., & Schendler, A. (2015). Science-based carbon targets for the corporate world: The ultimate sustainability commitment, or a costly distraction? *Journal of Industrial Ecology, 19*(6), 931–933.

Verrecchia, R. E. (2001). Essays on disclosure. *Journal of Accounting and Economics, 32*(1–3), 97–180.

Wahl, A., Charifzadeh, M., & Diefenbach, F. (2020). Voluntary adopters of integrated reporting–evidence on forecast accuracy and firm value. *Business Strategy and the Environment, 29*(6), 2542–2556.

Zhou, S., Simnett, R., & Green, W. (2017). Does integrated reporting matter to the capital market? *Abacus, 53*(1), 94–13.

# Conclusions

Corporations are increasingly called upon to contribute towards achieving a sustainable future. Moving away from a shareholder-oriented perspective, corporations are asked to bear greater responsibilities over their activities and impacts on the environment and society. Indeed, environmental, social, and corporate governance factors now represent a set of imperatives that corporate leadership must pursue to address sustainability priorities and objectives. Financing mechanisms have been put in place to support the sustainable development agenda, by integrating sustainability criteria into investment decisions to channel capital towards companies with responsible ESG practices, thus fostering a transition to a more sustainable economy. Investing in ESG-related initiatives holds the promise to offer long-term benefits in terms of company's improved economic performance and reduction in information asymmetry between the company and all stakeholders. Indeed, company managers have increasingly embraced the concept of corporate social responsibility and sustainability with the realization that they cannot only generate a successful strategy but can also be an effective means of cost control, for example, through reducing energy usage, waste flow, and purchasing low-impact resources.

The health care industry can play a pivotal role in the sustainability agenda. This industry can foster advancements in promoting good health and well-being, supporting equality, encouraging decent work and economic growth, and fostering partnerships. In the pursuit of a sustainable future, it is crucial for private businesses to step up by increasing transparency and showcasing their dedication to sustainability through thorough corporate reporting. This not only holds them accountable but also inspires others to follow suit, driving collective progress towards sustainability goals. The COVID-19 pandemic has profoundly impacted this industry, by leading to a surge in demand for health care services, straining resources and prompting a shift in priorities towards treating COVID-19 patients. This led to a significant increase in the use of digital health services, changing how routine care is provided while speeding up advancements in digital technologies. It has also



P. Candio, *Sustainability and Corporate Performance in Health Care*, CSR, Sustainability, Ethics & Governance,
https://doi.org/10.1007/978-3-031-63985-2

highlighted the importance of being adaptable, innovative, and cooperative in tackling future health emergencies, while also helping to progress towards sustainable objectives.

However, the health care industry faces significant challenges as regards its impact on the environment, specifically in terms of greenhouse emissions and solid waste. Digital health technologies are often touted as tools to lessen the environmental footprint of health care systems while maintaining service quality and financial stability. Nonetheless, the extent to which the implementation of these technologies can effectively address greenhouse gas emissions remains and other sustainability-related challenges uncertain in many cases. As remains uncertain whether digital health technologies in fact narrow, or widen the existing digital divide and health equity gap across the socioeconomic spectrum.

Sustainability reporting disclosure regulations and frameworks have evolved over time to serve as a mechanism for companies to fulfill regulatory obligations while also signaling their activities and commitments to the market and other stakeholders. In many jurisdictions including Europe, regulations mandate large companies to report on ESG performance, ensuring transparency and accountability. By disclosing information related to their sustainability practices, corporations can demonstrate compliance with these regulations, providing stakeholders with insights into their impact on the environment, society, and governance practices. This transparency fosters trust and confidence among stakeholders, influencing investment decisions and consumer choices. Companies that effectively disclose their sustainability efforts can differentiate themselves in the market, attracting socially responsible investors and customers who prioritize sustainability. Furthermore, by measuring and reporting on key ESG metrics, companies can benchmark their ESG performance against industry peers and track their sustainability performance over time. A number of ESG rating agencies today provide stakeholders and investors with information relevant to investment decision-making, though marked differences exist between industries, whereby the health care industry has its own peculiarities.

The pressure put on the top-ranking health care corporations listed in European stock exchanges has resulted in an overall improvement of their ESG profile over the last ten years. The implementation of the Directive 2014/95/EU, concerning non-financial reporting obligations, had minimal impact on the ESG performance of the sample companies, with marginal differences found with other industries suggesting that these companies had already adapted to forthcoming regulatory requirements due to the inherent significance of sustainability issues within the sector. The benchmarking analysis conducted revealed that the health care industry demonstrated significantly higher average profitability following the COVID-19 pandemic, unlike other industries which were adversely affected. This favorable performance of the health care sector amidst the pandemic suggests the essential nature of health care services and that the accelerated adoption of telehealth and digital health technologies during the pandemic may have enhanced operational efficiency and cost-effectiveness for health care companies.

The ESG performance of health care companies listed in the Euro STOXX 600 can modestly predict certain financial outcomes, namely, ROA and EBITM, though not EPS. Marked heterogeneity exists between the three ESG pillars however. The environmental score may have a positive association with ROA and ROE and so the social score with EPS. But the latter score can also negatively influence EBIT, and whereas the governance score predict positive EPS values only. This may have significant implications for policymakers and practitioners aiming to bolster ESG activities and non-financial reporting disclosures. The favorable influence of ESG activities on economic performance is likely to prompt boards of directors and management to evaluate the implementation of sustainability initiatives.

Diverse decisions concerning the choice of non-financial reporting framework, communication of social responsibility activities, and the activities and composition of listed companies have a significant, albeit heterogeneous, impact on analysts' ability to assess companies' future financial returns. Consequently, this variability affects information asymmetry between capital and stock markets and information intermediaries.

Future research studies should address the issue of causal inference in testing the influence that improvements in ESG profiles can have on corporate performance. Indeed, confounding factors that were not available for analysis might have masked reverse causality and more broadly unobserved heterogeneity issues. In addition, divergence between ESG scoring systems poses a major challenge in reliably drawing definitive conclusions about companies' sustainability profiles and their improvements over time. It is reasonable to expect that regulatory authorities will take steps to address these concerns and introduce regulatory safeguards for the utilization of ESG ratings in the foreseeable future. Embedding qualitative approaches in future quantitative examinations will enable a further examination of how investments in ESG-related activities are in fact detected by ESG rating agencies and perceived by stakeholders. Furthermore, the upcoming regulatory requirements for small and medium-size corporations demand careful investigation into how to integrate ESG considerations into business strategies and processes and ensure that the benefits reaped from such investments offset the initial significant costs.

# Appendix

Variables selected for analysis

| Variable | Code | Description |
|---|---|---|
| Punteggio totale ESG | esgscore | Overall company score based on self-reported information in the environmental, social, and governance sectors. |
| Punteggio pilastro ambientale E | envscore | Weighted average of the assessment assigned to a company based on reported environmental information and the three scores assigned to environmental categories. |
| Punteggio pilastro sociale S | socscore | Weighted average of the assessment assigned to a company, derived from reported social information and the four scores assigned to social categories. |
| Punteggio pilastro governance G | govscore | Weighted average of the assessment assigned to a company based on reported governance information and the three scores assigned to governance categories. |
| Forecast error | fe | Measure of information asymmetry obtained as the absolute difference between actual company earnings and analysts' forecasts for the year, subtracted from the average stock price. |
| Return on equity | roe | Return on equity, expressed as a percentage, calculated by subtracting preferred dividend requirements from net income and dividing the result by the average of the previous and current year's equity. |
| Return on asset | roa | Return on total assets, expressed as a percentage, takes into account net income, added interest expenses on debt deducted from capitalized interest, and the average of the previous and current year's total assets. |
| Earning per share | eps | Return level expressed in terms of earnings per share. |
| Price earning ratio | per | Average stock price divided by the earnings per share rate at the requested date. |
| Total asset | totass | Represents the sum of total current assets, long-term receivables, investments in unconsolidated subsidiaries, other investments, net property, plant, and equipment, and other assets. |



P. Candio, *Sustainability and Corporate Performance in Health Care*, CSR, Sustainability, Ethics & Governance,
https://doi.org/10.1007/978-3-031-63985-2

| Variable | Code | Description |
|---|---|---|
| Book value per share | bvps | Represents the book value (common equity divided by outstanding shares) at the end of the fiscal year. |
| Market to book value | mvbv | Market value of common equity divided by the reported common equity value in the company's balance sheet. |
| Market capitalization | mc | Market value obtained by multiplying the stock price by the number of outstanding common shares. |
| Leverage | lev | Total debt divided by the total resources available to the company. |
| Net income | netinc | Annual profit excluding extraordinary activities. |
| Earnings before interest and tax | ebit | Income of a company before the application of interest and income taxes. It is calculated by taking pre-tax income and adding debt interest expenses and subtracting capitalized interest. |
| Volatility | beta | Measure of volatility, or systematic risk, of a security or portfolio relative to the market as a whole. A beta greater than 1 indicates higher volatility compared to the market. |
| Number of estimates | epsy1 | Total number of financial estimates made by financial analysts. |
| Market price | mrkprice | Average stock price at the end of the relevant month. |
| Sustainability Committee | csrsc | Answers the question: Does the company have a CSR committee or team for corporate social responsibility? |
| Sustainability External Audit | csrsea | Answers the question: Does the company have an external auditor for its CSR, safety, and sustainability report? |
| ESG Reporting Scope | csresgrs | Percentage of the company's activities included in its environmental and social report. |
| Sustainability Reporting | csrsr | Answers the question: Does the company publish a separate report on corporate social responsibility, safety, and sustainability, or does it publish a dedicated section within its annual CSR, safety, and sustainability report? |
| Stakeholder Engagement | csrse | Answers the question: Does the company communicate how it engages with various stakeholders? |
| Integrated Strategy | csria | Answers the question: Does the company explicitly integrate financial and extra-financial factors in the management analysis and discussion section (MD&A) in the annual report? |
| Global Reporting Initiative | gri | Answers the question: Has the company's corporate social responsibility report been published in accordance with the Global Reporting Initiative guidelines? |
| Industry | icbic | Company's sector of activity. |

# GPSR Compliance

*The European Union's (EU) General Product Safety Regulation (GPSR) is a set of rules that requires consumer products to be safe and our obligations to ensure this.*

*If you have any concerns about our products, you can contact us on ProductSafety@springernature.com*

In case Publisher is established outside the EU, the EU authorized representative is:

Springer Nature Customer Service Center GmbH
Europaplatz 3
69115 Heidelberg, Germany

The manufacturer's authorised representative in the EU is Springer Nature Customer Service Centre GmbH, Europaplatz 3, 69115 Heidelberg, Germany. If you have any concerns regarding our products, please contact ProductSafety@springernature.com

Printed and bound by CPI Group (UK) Ltd, Croydon, CR0 4YY
01/10/2026
02225977-0010